ILLUSTRATED ATLAS Of MUSCULOSKELETAL ANATOMY

The *Illustrated Atlas of Musculoskeletal Anatomy* is intended to be a beneficial resource for use in both education and clinical practice. The author has made an earnest effort to present, in a clear and precise fashion, the most accurate information available. Numerous reference sources were consulted in the preparation of both the illustrations and text, to insure that material presented herein is reliable and valid.

Knowing that this work will not be found without fault, the author and publishers invite the candid criticism of those who would pay us the honor of reading these pages, trusting that the author will have the privilege of correcting, in future editions, any errors that might have escaped his attention.

1st Edition - 2005 **ISBN 097119262-6** Copyright Pine Island Publishers, Inc.

For information on this and other publications please contact:

Pine Island Publishers, Inc.
Phone: **1-800-400-1892**

www.PineIslandPublishers.com
www.ExamReviews.com

ACKNOWLEDGMENTS

The author wishes to extend his deepest appreciation to all those who have played a role in bringing this text to fruition. I especially thank Leigh Reynolds of Pine Island Publishers, Inc. for her many hours of editing, technical contributions, and personal support. A special thanks to Jeff Weaver and Ellen Caravatti for reviewing the book. My thoughts go to all my students, teachers, and colleagues who continue to educate me. To my mother who encouraged me to read, to question, and to think, I am forever grateful.

Cover design & layout by Christine Reuter.

Anatomical illustrations by Dianne Beneditti and Sandy Golding.

Watermark images are from the sketchbook of Leonardo da Vinci.

Musculoskeletal illustration on front cover by the author.

ABOUT THE AUTHOR

Dr. Patrick Barron received both his Bachelor's and Master's Degrees in Rehabilitation from Marshall University in Huntington, WV. His Medical Degree was earned at Bastyr University in Seattle, WA. He completed two years Post Doctoral training at the University of Washington and has been in private practice specializing in Rehabilitation Medicine since 1984. Dr Barron is the director of Oviedo Physical Medicine, a medical and educational consulting group, and is the author of numerous health related books and published articles. He is the founder of the Central Florida School of Health Sciences and currently teaches at several Florida Colleges.

TABLE of CONTENTS

BRIEF HISTORY OF ANATOMY

The oldest known anatomical treatise is an Egyptian papyrus probably written around 1600 BC. It shows the heart, vessels, liver, spleen, kidneys, ureters, and bladder were recognized, and certain blood vessels were known to come from the heart. Other vessels are described, some carrying air, some mucus, while two to the right ear are said to carry the breath of life, and two to the left ear, the breath of death. **Imhotep**, in the 3rd dynasty, is credited as the founder of ancient Egyptian medicine and as the original author of the Edwin Smith Papyrus. This papyrus included cures and ailments as well as anatomical observations. The Edwin Smith Papyrus is regarded as a copy of numerous earlier works. Some information contained in it actually dates from 3000 BC.

Aristotle (c384-322 BC) is primarily known for his writings on philosophy, but is believed to be the first anatomist of the modern western world. Born in northern Greece, his father was physician to the royal family. Aristotle first studied philosophy with Plato in Athens and later astronomy and medicine in Asia Minor. Though his work in zoology is not without errors, he is known to have written with great anatomical detail concerning nerves, tendons, and blood vessels. His descriptions of animal anatomy and his taxonomic classifications are still remarkably modern sounding.

Hippocrates, of the island of Cos (320 – 260 BC), is universally considered to be the "Father of Medicine" and has also been thought to be the creator of the science of anatomy. It does appear, from his few documented personal writings, that Hippocrates had some accurate notions on osteology, but in general his ideas on the structure of the human body were both quite superficial and invalid. Of the muscles, soft tissues in general, and the internal organs, his ideas are confused, indistinct, and erroneous. He seems to have been unaware of the distinction between vein, artery, nerve, or tendon. Hippocrates did, however, reject the superstitious views of his day and base his medical beliefs upon observation and study of the human body, stating that illness had a physical and rational explanation. He taught that the four elements (earth, air, fire, and water) were analogous to the four constituent elements in man (phlegm, blood, yellow bile, black bile). Although scientifically worthless as a theory from our modern viewpoint, his teachings remained the foundation of European medical science for more than two millennia.

Erasistratus and **Herophilus** (c300 BC) perform the first human dissections in Greece.

Claudius Galen (c129-201 AD), a Greek physician from Pergamum who settled in Rome, was the first anatomist to begin recording a complete picture of the tissue lying beneath the skin. Galen used his position, as physician to the gladiators to further his knowledge of human anatomy. Because human dissection was illegal, much of Galen's anatomical research was limited to animal studies. Although now seen as deeply flawed, Galen's very detailed and organized writings on anatomy and medicine were well accepted in Europe for over 1,300 years.

Mondino dei Luzzi (1275-1326), professor of Medicine at the University of Bologna, wrote the first treatise on anatomy based upon the dissection of cadavers. This manuscript and its artwork began the trend of using anatomical illustrations in textbooks, which would become the standard in the following centuries. **Guido da Vigevano** (1280-1349), a contemporary of Mondino dei Luzzi, contributed significantly to the understanding of neuroanatomy and was equally influential in breaking free from the dominating influence of Galen.

Leonardo da Vinci (1452-1519) was the first anatomist-artist to accurately draw anatomical structures of the human face and hands, thus laying the foundation for all modern scientific illustration. By his own account, da Vinci dissected more than 30 human cadavers, first in Milan and later in the hospitals of Florence and Rome. Da Vinci's *Virtuvian Man*, which shows the male human form as it relates to the circle and the "golden ratio", is one of the most well known of anatomical drawings. He ultimately saw anatomy in an emotional rather than a purely scientific way, and therefore, some of his later illustrations reflect his anatomically flawed beliefs.

Michelangelo Buonarroti (1475-1564), Italian sculptor, painter, architect, and poet is renowned as one of the three greatest artists of the Renaissance. It is well known that Michelangelo did "flay dead bodies in order to discover the secrets of anatomy". His sculpture *David* and painting the *Creation of Adam* reflect his amazing knowledge of human anatomy. In fact, Frank Lynn Meshberger, M.D., reported in the October 1990 JAMA, that Michelangelo's *Creation of Adam* contains the image of God superimposed upon an amazingly (anatomically) accurate and detailed image of the human brain.

Andreas Vesalius (1514-1564), a Flemish physician and teacher, had possibly the greatest influence on the development of modern anatomy. Vesalius, aided by a softening in the position of the church on dissection, performed numerous detailed dissections, layer-by-layer, recording data about the human anatomical structures. In 1543, at the age of 28, he teamed with Renaissance artists to produce his *De Humani Corporis Fabricia*, the first comprehensive textbook of anatomy. Despite the cry of outrage that arose from the present day scientific community, Vesalius persisted and took his theatrical anatomy show on the road. In 1555, he published a second edition of the *De Fabricia*, which was ultimately accepted by his peers.

Peter Paul Rubens (1577-1640), the Flemish Baroque painter, fused the realism of traditional Flemish portrait style with the imaginative classical themes of the Italian Renaissance. Rubens imbued many of his anatomically detailed paintings with glowing color and flickering light to inject an exuberance not usually found in previous anatomically based art.

Fredrick Ruysch (1638-1731), a Dutch physician and professor, became the first great proponent of the "anatomical specimen". He used the cadavers of stillborns and dead infants to create extraordinary multi-specimen scenes. Ruysch claimed the right to collect and exhibit human material without the consent of the subjects.

John Browne (1642-1702), a British anatomist and author, published the *Compleat Treatise of the Muscles* in 1681. His text was reprinted four times, yet is also subject to accusations of plagiarism, as it bares great resemblance to the works of **Guilio Casserio**. The anatomical subjects in the work of these anatomists seem to dance and posture with over-the-top theatrical gestures, transforming the lowly cadaver into a highborn performing socialite.

Govard Bidloo (1649-1713), a Dutch anatomist, took anatomical illustration in a new direction. His work included, not only the empirical anatomical structures, but also the prosthetics of dissection (ropes, props, nails, chains), as well as the mutilation and ugliness of the dissected body. Bidloo appropriated and dissected cadavers with an amazing clinical detachment.

The **Anatomy Act** (1832**)** was passed in Britain, allowing "unclaimed" pauper bodies to be sold or appropriated for use in dissection.

Frank Netter (1906-1991), a successful American commercial artist for such publications as *The Saturday Evening Post* and the *New York Times*, later entered medical school and became the most well known anatomical illustrator-physician of our time. Netter's finest works include the 13 volume *Ciba Collection of Medical Illustrations* and his *Atlas of Human Anatomy*. Netter stated that he attempted to depict realistic living subjects whenever possible, as those were the primary interest of most physicians.

Gunther von Hagens, (1945-), a Polish born physician and artist, completed his medical studies at the University of Heidelberg in 1973 and invented "Plastination" in 1977. Plastination is a process that replaces the natural body fluids and fats with synthetic polymers, and preserves the cadaver indefinitely. Dr. von Hagens has been displaying his work since the late 1980's. Over 8 million people have seen his anatomical exposition "Bodyworks" since opening in 1995. Unlike plastic anatomical models, plastinated specimens are intricate displays of "real" human anatomy. Dr. von Hagens plans to be plastinated when he dies.

Damien Hirst (1965-), a British born artist, produced and self-promoted his *Freeze* and *Natural History* exhibitions, which have included dead animals and fish suspended in tanks of formaldehyde. These and other works like *A Thousand Years*, which include the carcasses of rotting animals encased in glass, are designed not for their anatomical presentation, but more for their ability to provoke "thought" and "fright".

Anthony Noel Kelly, a British butcher and laborer-turned-artist, shocked the art world in 1997 when he exhibited a series of plaster casts of cadavers, painted silver and pinned to a wall. Kelly, who had been given permission to sketch the cadavers by the Royal College of Surgeons, was subsequently arrested and sentenced to nine months in jail for stealing the bodies. Kelly describes himself as" being detached from death" and states that working with dead things is therapeutic.

Human Anatomy

Body Planes

When studying the human body, it is necessary to divide it into spatial planes of reference. The standard terms for these planes are listed below.

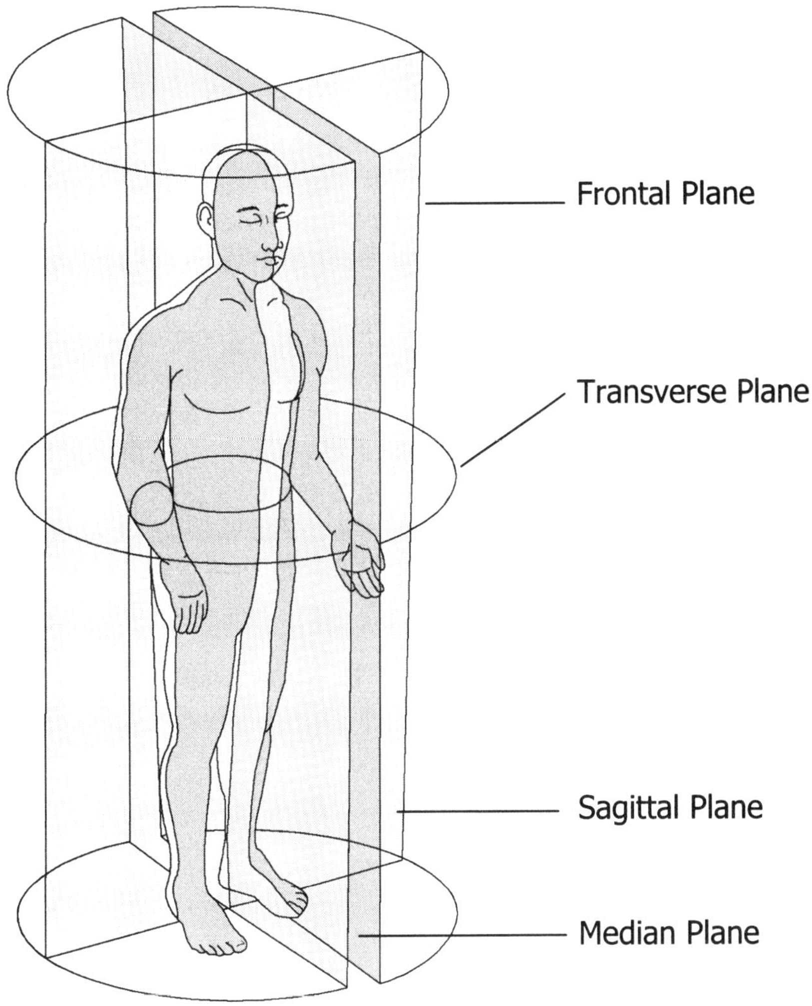

Frontal or coronal -- a lengthwise plane that divides the body or any of its parts into anterior and posterior portions.

Transverse or Horizontal -- a crosswise plane that divides the body or any of its parts into upper and lower parts.

Median -- a lengthwise plane through the midline that divides the body or any of its parts into equal left and right sides.

Sagittal -- a lengthwise plane running from front to back that divides the body or any of its parts into left and right sides. A *midsagittal* plane is equivalent to a median plane.

Anatomical Directional Terms

Anatomical Position -- position of reference in which the subject is in a standing position, with feet together and palm of the hands forward

Superior, Cranial or Cephalic -- toward the head or upper end of the body (ex: the heart is superior to the stomach)

Inferior or Caudal -- away from the head or toward the lower (tail) end of the body *(ex: the pelvis is inferior to the chest)*

Anterior or Ventral -- front or more forward than another structure (ex: the sternum forms the anterior boundary of the chest)

Posterior or dorsal -- back or more behind than another structure (ex: the spine forms the posterior boundary of the chest)

Medial -- toward the midline of the body, extending toward the midline (ex: the great toe is located at the medial side of the foot)

Lateral -- away from the midline of the body, extending toward the side (ex: the ear is located on the lateral portion of the head)

Proximal -- toward the point of origin of a part of the body, nearer to the trunk (ex: the elbow is proximal to the hand)

Distal -- away from the point of origin of a part of the body, farthest from the trunk *(ex: the foot is distal to the knee)*

Introduction To The Human Skeleton

The Human skeleton is divided into two major systems of bones: the axial skeleton and the appendicular skeleton. Together they contain 206 named bones.

The *axial skeleton* consists of 80 bones including 74 that form the upright axis of the body and 6 in the middle ear (the osicles: malleus, incus & stapes).

Axial Skeleton: (80 bones)

1. Skull (8 cranium, 14 face, 6 middle ear)
2. Vertebrae (24)
3. Sacrum (fused in the adult, from 5 vertebral segments)
4. Coccyx (fused in the adult, from 4 or 5 vertebral segments)
5. Sternum (1, fused from the Manubrium, Body & Xiphoid Process)
6. Ribs (12 pairs)
7. Hyoid bone (1)

The *appendicular skeleton* contains 126 bones and forms the free moving appendages that attach to the axial skeleton. These bones make up the shoulder girdles, arms, wrists, hands, hip girdles, legs, ankles, and feet. They are further categorized as the upper appendicular and lower appendicular skeletons. The skeletal bones are listed as:

Upper Appendicular Skeleton (64 bones)

1. Clavicle (2)
2. Scapula (2)
3. Humerus (2)
4. Radius (2)
5. Ulna (2)
6. Carpals (8 pairs)
7. Metacarpals (5 pairs)
8. Phalanges (3 in digits 2 through 5, 2 in each thumb)

Lower Appendicular Skeleton (62 bones)

1. Pelvis (2 bones in adult, from the fused: ilium, ischium & pubis)
2. Femur (2)
3. Patella (2)
4. Tibia (2)
5. Fibula (2)

6. Tarsals (7 pairs)
7. Metatarsals (5 pairs)
8. Phalanges (3 in digits 2 through 5, 2 in each great toe)

Bone Markings

Different points on bones are labeled by the depressions and projections of their structures. This helps when locating other structures such as muscles, nerves, and blood vessels. Below is a list of some of the more common terms for these structures.

Depressions

Fossa -- a depression or hollow
Foramen -- a hole, perforation or orifice
Meatus -- a tube-shaped opening or passage
Sinus -- a cavity or hollow in a bone

Projections

Condyle -- a rounded projection
Crest -- a prominent ridge
Epicondyle -- a projection at the end of a bone, above the condyle
Facet -- a small plane surface
Head -- a rounded projection beyond a narrow neck like portion
Line -- a narrow ridge
Spinous Process or spine -- a sharp projection
Trochanter -- a very large projection
Tubercle -- a small rounded projection
Tuberosity -- a large projection

Classification of Bones

Bones are classified as being *long*, *short*, *flat*, *irregular*, or *sesamoid* in shape.

1. *Long* bones comprise most bones of the extremities, ranging from the shorter phalanges to the longest bone, the femur. The shaft (or diaphysis) of a long bone encloses an elongated marrow cavity. The extremities, or ends, of long bones are called epiphyses.
2. *Short* bones have approximately the same dimensions in all directions. The cube-like bones of the wrist (carpals) and ankles (tarsals) are examples of short bones.
3. *Flat* bones are those with two surfaces roughly parallel and close to each other, as in the sternum, the scapula, and most bones of the skull.
4. *Irregular* bones fit the description of none of the above categories. Vertebrae are examples of irregular bones.

5. *Sesamoid* bones are shaped like sesame seeds. They are found within tendons and act like pulleys to insure smooth movement. The patella is a sesamoid bone, others are found in the hands and feet.

Classification of Joints

Joints between bones are called articulations. These articulations hold bones firmly together and yet permit movement between the bones [Greek, *arthro* meaning joint] There are three main types of joints:

1. *Synarthroses* or *Fibrous joints* -- immovable (or least moveable) joints
2. *Amphiarthroses* or *Cartilaginous joints* -- slightly movable joints
3. *Diarthroses* or *Synovial joints* -- freely movable joints

Synarthroses

These joints have **fibrous** tissue growing between the articulation surfaces of the bones. This tissue binds the bones so closely together that movement cannot occur. [Greek, *syn* meaning with] There are three types of synarthrotic joints:

1. *Sutures* -- only found in the skull, these tooth-like projections from adjacent cranial bones are connected by fibrous, ligament-like tissue
2. *Syndesmosis* -- joints in which bony surfaces are bound together by fibrous interosseous ligaments or membranes (interosseous ligament between the radius and ulna & tibia and fibula).
3. *Gomphosis* -- peg-and-socket joint between the root of a tooth and the jawbone, bound by periodontal ligament

Amphiarthroses

This type of joint is comprised of bones connected by **cartilage**. The cartilage allows only slight movement between the bones. [Greek, *amphi* meaning both] There are two types of amphiarthrotic joints:

1. *Symphysis* -- joint in which two bones are separated by a disk of fibrocartilage (pubic symphysis)
2. *Synchondrosis* -- joint in which two bones are connected by hyaline cartilage (between the ribs and sternum)

Diarthroses

These **synovial** joints are characterized by a space separating the articulating surfaces of the bones. [Greek, *di* meaning double or twice] They are the most mobile of the joints, as well as the most numerous and complex. All diarthrotic joints consist of a joint capsule, synovial membrane that produces synovial fluid, and both intrinsic and extrinsic ligaments. Each approximating bony end is covered with articular (hyaline) cartilage.
There are three types of diarthrotic joints and each type is divided into two subtypes:

1. *Uniaxial:* permit movement in only one plane and around one axis.
   a. *Hinge joints* -- permit only back and forth movement (humeroulnar, femurotibial, and interphalangeal)
   b. *Pivot joints* -- permit rotation of a projection of one bone within the notch or ring of another bone (atlanto-axial, radioulnar)

2. *Biaxial:* permit movement around two perpendicular axes in two perpendicular planes.
   a. *Saddle joints* -- articulating ends of the bones resemble reciprocally shaped miniature saddles (carpal-metacarpal joints of thumbs)
   b. *Gliding joints* -- flat surfaces allow gliding movement (intercarpal joints and facet-facet joints of vertebrae)

3. *Multiaxial:* permit movement around three or more axes and in three or more planes.
   a. *Ball & socket joints* -- round end of one bone fits into the cavity of another bone. The most mobile of all joints - (glenohumeral)
   b. *Condyloid joints* -- permits all forms of angular movement except axial rotation (radiocarpal)

Typical Joints

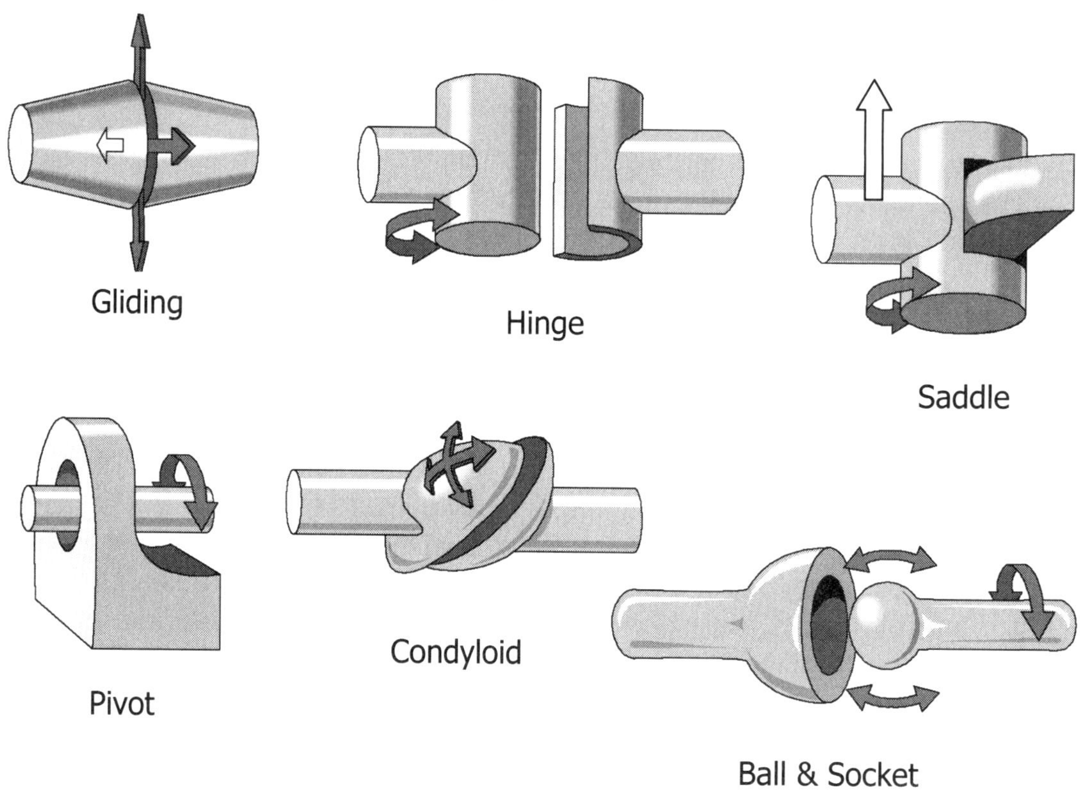

Introduction To Muscles of the Human Body

Muscles make bones and joints move. The nearly 700 skeletal muscles comprise up to 40-50% of the adult human body (by weight), and function to provide us with movement, heat production, and posture.

Muscle Fiber Patterns

Fibers within skeletal muscle form patterns, which are visible to the naked eye. When the fibers are aligned, such as in the abdominal muscles, the pattern is described as parallel or **strap**-like (quadratus lumborum & rectus abdominis). Where the fibers run obliquely to the direction of force, the muscle is described as either **triangular** or **pennate**. If the muscle fibers insert diagonally into a tendon that runs the entire length of the muscle, the muscle is described as pennate. **Bipennate** muscles (rectus femoris) have their fibers inserting obliquely on both sides of the tendon. In **fusiform** muscles (biceps brachii), the fibers may be nearly parallel in the belly, yet converge to fuse at a tendon at one or both ends. **Multipennate** muscles (deltoid) have multiple oblique fiber directions. Some muscles have a **spiral** or twisted arrangement (latissimus dorsi).

Strap

Pennate

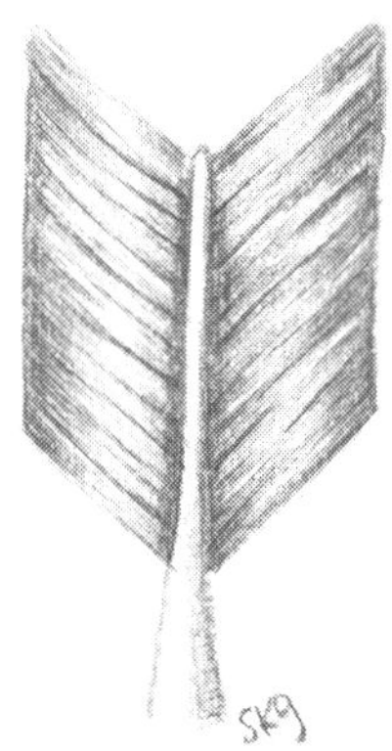

Bipennate

Fusiform

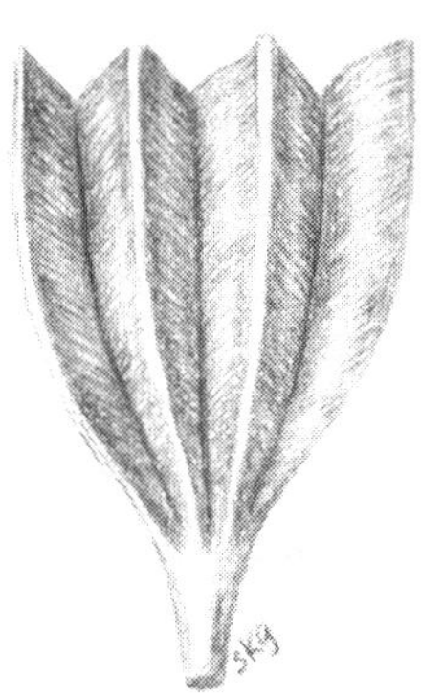

Multipennate

Triangular

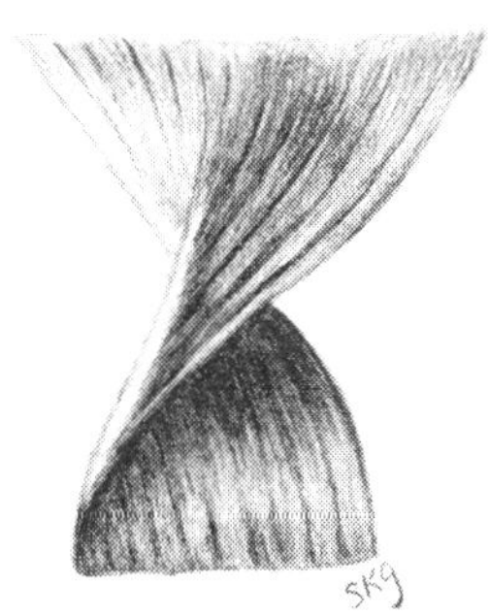

Spiral

Tricipital

Action of Muscles

The action of a muscle is defined as the movement that takes place when that muscle contracts. In this context, contraction is equated to shortening and relaxation is equated to lengthening. Contraction may be considered as an attempt on the part of the muscle to bring its attachments closer together. The success of a muscle to accomplish this task depends upon a number of factors including gravity, inertia, external loads, opposing muscles, and forces generated by the elastic and viscous properties of associated structures. Depending on circumstances, increasing muscle tension may result in shortening, lengthening, or maintaining its original dimensions. The three major types of muscle contractions are **isotonic**, **tonic**, and **isometric**.

1. **Isotonic** contractions result in a change of the active muscle length
   Concentric isotonic contractions result in the shortening of the active muscle - positive work (up-phase of a bicep curl)
   Eccentric isotonic contractions occur as there is lengthening of the active muscle - negative work (controlled down-phase of a bicep curl)

2. **Tonic** contractions involve increased muscle tension, but do not change the active muscle length and are used to maintain position or posture

3. **Isometric** (static) contractions involve increased tension in the active muscle, but overpowering external forces prevent movement

Movement

Natural movements are accomplished by using groups of muscles. Muscles are classified with regard to the role played during a particular movement. **Prime movers** (agonists) are characterized as those muscles that consistently initiate and maintain a particular movement. **Antagonists** are characterized as those muscles that wholly oppose a movement or consistently initiate and maintain the opposite movement. When prime movers and antagonists contract together they act as **fixators**. Muscles that assist in accomplishing a movement are known as **synergists**. Given different situations, a particular muscle may act as a prime mover, antagonist, fixators, or synergist. Muscles perform work in a variety of situations. Convention and habit dictate that the more stable (usually heavier and more proximal) attachment is referred to as the **origin**. The more mobile (usually lighter and more distal) attachment is referred to as the **insertion**. The relative mobility of bony attachments may necessitate that use of these two terms be closely considered with regard to any given situation.

Biomechanics

Movement is brought about by the action of muscle upon rigid skeletal components or levers. A lever is defined as a rigid bar that turns about an axis or fulcrum. Bones provide this mechanical advantage (levers) by which work may be accomplished. Joints form the pivotal point of movement (fulcrum) around which the levers move. By changing the location of the fulcrum with regard to the force or effort (muscle) and resistance (weight), it is possible to move heavier loads and to alter both the rate and the distance over which the load can be moved.

There are three classes of levers:

In **class I** levers, the fulcrum is located between the point at which the force is applied and the weight that is to be moved. A teeter-totter is a common example of a class I lever. In the human body, this type of lever is involved in extending the elbow against resistance.

In **class II** levers, the weight to be moved is located between the fulcrum and the point of force. A wheelbarrow typifies this type of lever. In the human body, this type of lever is involved in rising up onto the toes via plantar flexion. There are relatively few occurrences of class II levers in the human body.

In **class III** levers, the weight is on one end, the fulcrum at the other, and the force is applied between them. The lifting of a shovel utilizes this type of lever. Class III levers are the most common type in the human body. One example is flexion of the elbow.

In the human body, class I levers provide for increased range of motion, with a reduction in available strength. Class II levers provide for increased strength with a reduction of range of motion. Class III levers provide for an increase in range of motion, but a reduction of available strength.

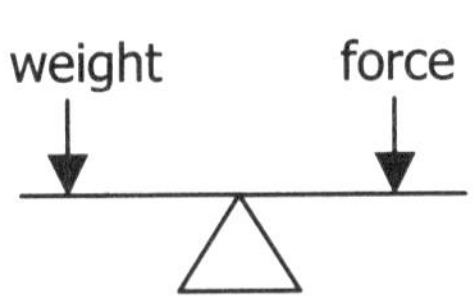

Class I Lever

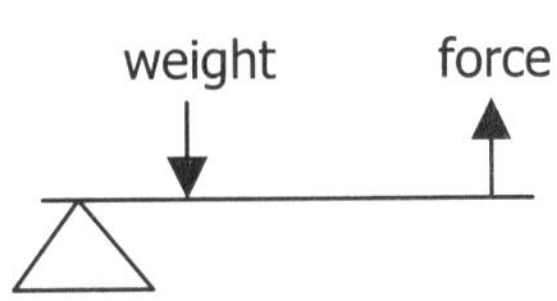

Class II Lever

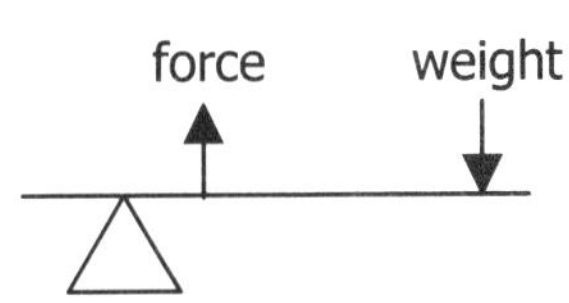

Class III Lever

Movement Terminology

Abduction -- movement away from the medial plane of the body
Adduction -- movement toward the medial plane of the body (retraction)
Circumduction -- movement of a part so its distal end describes a circle and the rest of the part describes a cone
Depression -- movement of a part downward
Dorsiflexion -- raising the foot upward
Elevation -- movement of a part upward
Eversion -- turning the sole of the foot outward (also described as pronation)
Extension -- increasing the angle between bones
Flexion -- decreasing the angle between bones
Hyperextension -- movement of an extended part beyond is normal anatomical position
Inversion -- turning the sole of the foot inward (also described as supination)
Lateral rotation -- rotation away from medial plane (also described as external rotation)
Medial rotation -- rotation toward the medial plane (also described as internal rotation)
Opposition -- movement of thumb and little finger toward midline of hand
Plantarflexion -- extending the foot downward (pointing the toe downward)
Pronation -- rotating the forearm inward & the back of the hand forward (pronation of the foot & ankle involves eversion, abduction, and external rotation)
Prone -- position with face down or dorsal surface up
Protraction -- movement of a part forward
Retraction -- movement of a part backward
Rotation -- pivoting a bone upon its own axis
Scapular Abduction -- movement of scapula away for the medial plane of the body (protraction)
Scapular Adduction -- movement of scapula toward the medial plane of the body
Supination -- rotating the forearm outward, turning the palm of the hand forward as in anatomical position
Supine -- position with face or ventral surface upward

Muscle Terminology

The names given to muscles are usually descriptive, based upon their relative shapes, sizes, attachments, actions, numbers of heads, positions, or depths. Below are descriptive anatomical terms used for muscles, locations, or actions.

Abdominis - - the abdomen
Anterior -- front or ventral (with humans in anatomical position)

Biceps -- two heads
Brachii -- arm
Brevis -- short
Carpi -- related to the carpals or wrist
Deltoid -- triangle
Digastric -- two bellies
Digiti minimi - - little finger
Digitorum -- finger or toes
Dorsi -- back
Extensor -- producing the action of extension
External -- to the outside
Flexor -- producing the action of flexion
Gemellus -- twins
Gracilis -- slender
Hallucis -- great toe
Indices -- index finger
Infraspinatus -- below the spine
Internal -- to the inside
Interosseus -- between bones
Latissimus -- broadest
Longissimus -- longest
Longus -- long
Major -- greater
Minor -- lesser
Obturator -- plugged
Palmaris -- palm
Pectoralis -- the chest
Peroneus -- related to the fibula
Pollicis -- thumb
Posterior -- back or dorsal (with humans in anatomical position)
Profundus -- deep
Quadratus -- square
Quadriceps -- four heads
Radius -- related to the radius bone
Rectus -- straight
Rhomboid -- diamond shaped
Splenius -- bandage
Superficialis -- closer to the surface
Supraspinatus -- above the spine
Teres -- round
Tibialis -- related to the tibia
Triceps -- three heads
Ulnaris -- related to the ulna bone
Vastus -- immense

Kinesiology & Gait

Gait is defined as the manner or style of locomotion (walking, running, etc.) and includes the components of rhythm, cadence, and speed.

A ***gait cycle*** is described as being the period in which a complete sequence of events takes place, beginning with the heel of one foot striking the floor and ending with the heel of the other foot striking the floor. The gait cycle is divided into two categories of events: the *stance phase* and the *swing phase.* The stance phase is defined as the time period when the limb being considered is in contact with the floor. The swing phase is defined as the time period when the limb being considered is not in contact with the ground. In walking, there always exists a period of time when both feet are in contact with the ground. This portion of the gait cycle is known as *double stance*. In the average walking person, a stance phase occurs during approximately 60% of the gait cycle while the swing phase occurs during approximately 40% of the gait cycle. As the speed of walking increases, the time spent in stance phase diminishes. Running is defined, as the gait in which there exists no double stance phase (i.e. at any given time, only one foot is in contact with the ground).

In humans, gait is primarily determined by the pelvis, upper leg (thigh), lower leg and the foot. The upper body components including the trunk, head, and arms are used in locomotion to provide momentum and counterbalance.

The components of the stance phase of gait include: ***heel strike*** (hip flexed with knee & ankle in neutral), ***foot flat*** (hip flexed with knee & ankle flexed), ***midstance*** (hip, knee & ankle in neutral), ***heel-off*** (hip extended, knee in neutral & ankle dorsiflexed), and ***toe-off*** (hip in neutral with knee flexed & ankle dorsiflexed).

The components of the swing phase of gait include: ***acceleration*** (hip flexed with knee flexed & ankle in plantar flexed), ***midswing*** (hip flexed, knee flexed & ankle in neutral), ***deceleration*** (hip flexed with knee & ankle in neutral) and ***arm swing*** (contralateral arm & leg moving in the same direction in the sagittal plane -- as each shoulder girdle advances, the pelvis & lower limb of the ipsilateral side trail behind).

The *main determinants* of gait have been defined as: *pelvic rotation, pelvic tilt, knee & hip flexion, knee & ankle interaction*, and *lateral pelvic displacement*.

It is during the *midstance* phase that the *center of gravity* of the human body reaches it's highest point in the gait cycle.

The Skeletal System

The human skeletal system is composed of several different types of tissues working together: osseous tissue (bone), cartilage, dense fibrous connective tissue (ligament), adipose tissue, circulatory tissue (blood vessels), and nervous tissue. This combination represents a complex, dynamic, living structure. Skeletal tissue is continuously changing or remodeling itself (via: resorption & deposition) to meet the demands of growth, exercise, and repair. A process known as *ossification* or *osteogenesis* forms bones. Mature bone cells are known as *osteocytes*. Cartilage cells are known as *chondrocytes*. Mineral depositing cells are known as *osteoblasts*, while mineral resorbing cells are known as *osteoclasts*. It is through the action of these cells that bone remodeling occurs. Bone growth and repair depend on a number of factors including hormone levels, nutrient availability, and mechanical stress (weight bearing exercise).

Bone (osseous) tissue contains a significant amount of non-living intercellular tissue known as matrix. This matrix is composed of approximately 25% water, 25% collagen fibers, and 50% crystallized mineral salts. The most abundant mineral salt is known as hydroxyapatite, which contains calcium phosphate and calcium carbonate. Other mineral salts commonly found in bone are: magnesium hydroxide, fluoride, and sulfate. It is the deposition of these minerals in the collagen fiber network that gives bone its hardness.

Skeletal tissue provides support and protection, assists in the production of blood cells, and stores minerals and triglycerides. For the purposes of this text, we will focus our attention on the structural attributes of the skeletal system. These structures provide a framework on which skeletal muscles may be attached and which allow for movement and flexibility.

The following illustrations of the major skeletal structures and associated joints are labeled to make obvious only their prominent bony landmarks and most significant anatomical or functional features.

SKELETON
Ventral (Anterior) View

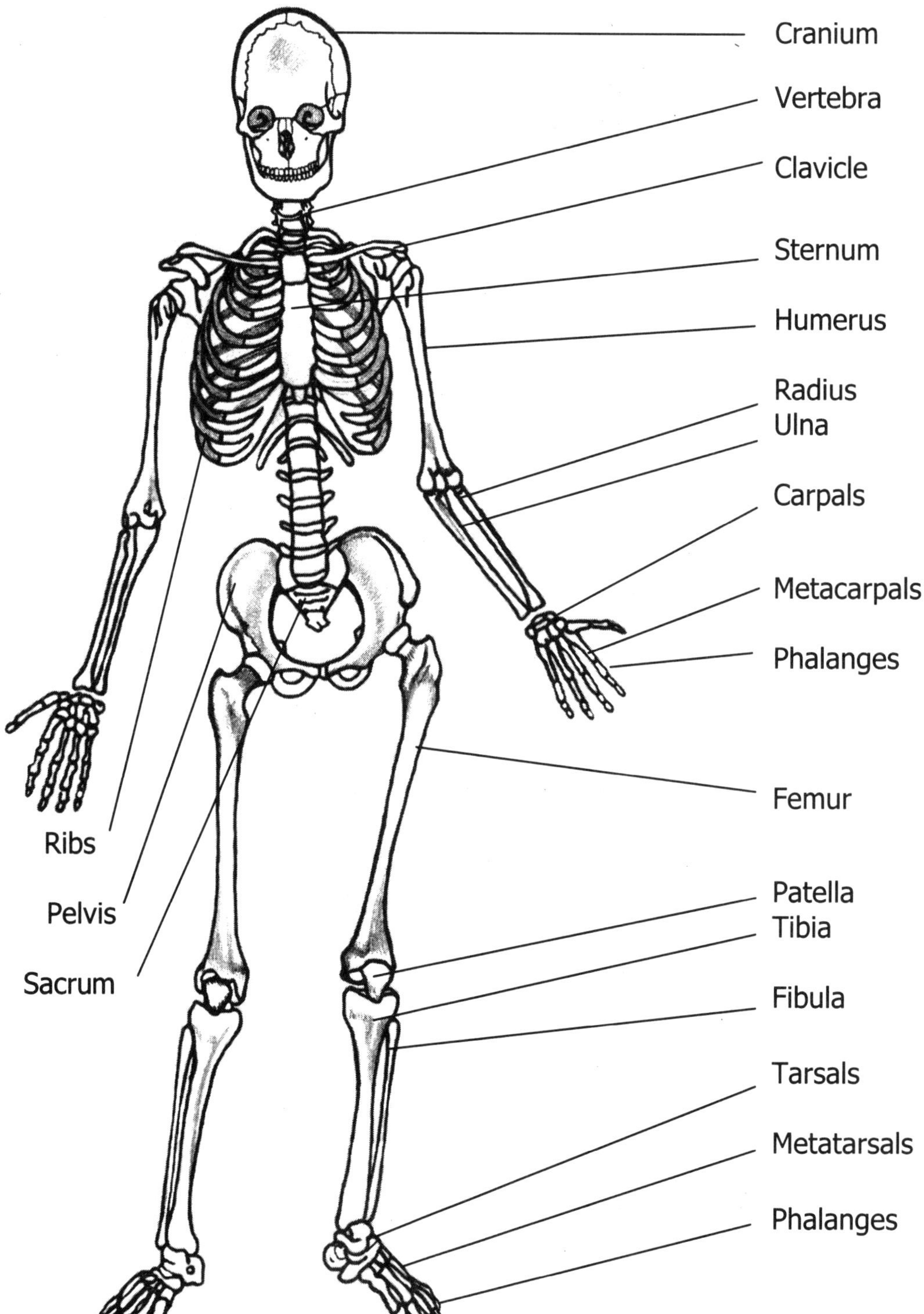

SKELETON
Dorsal (Posterior) View

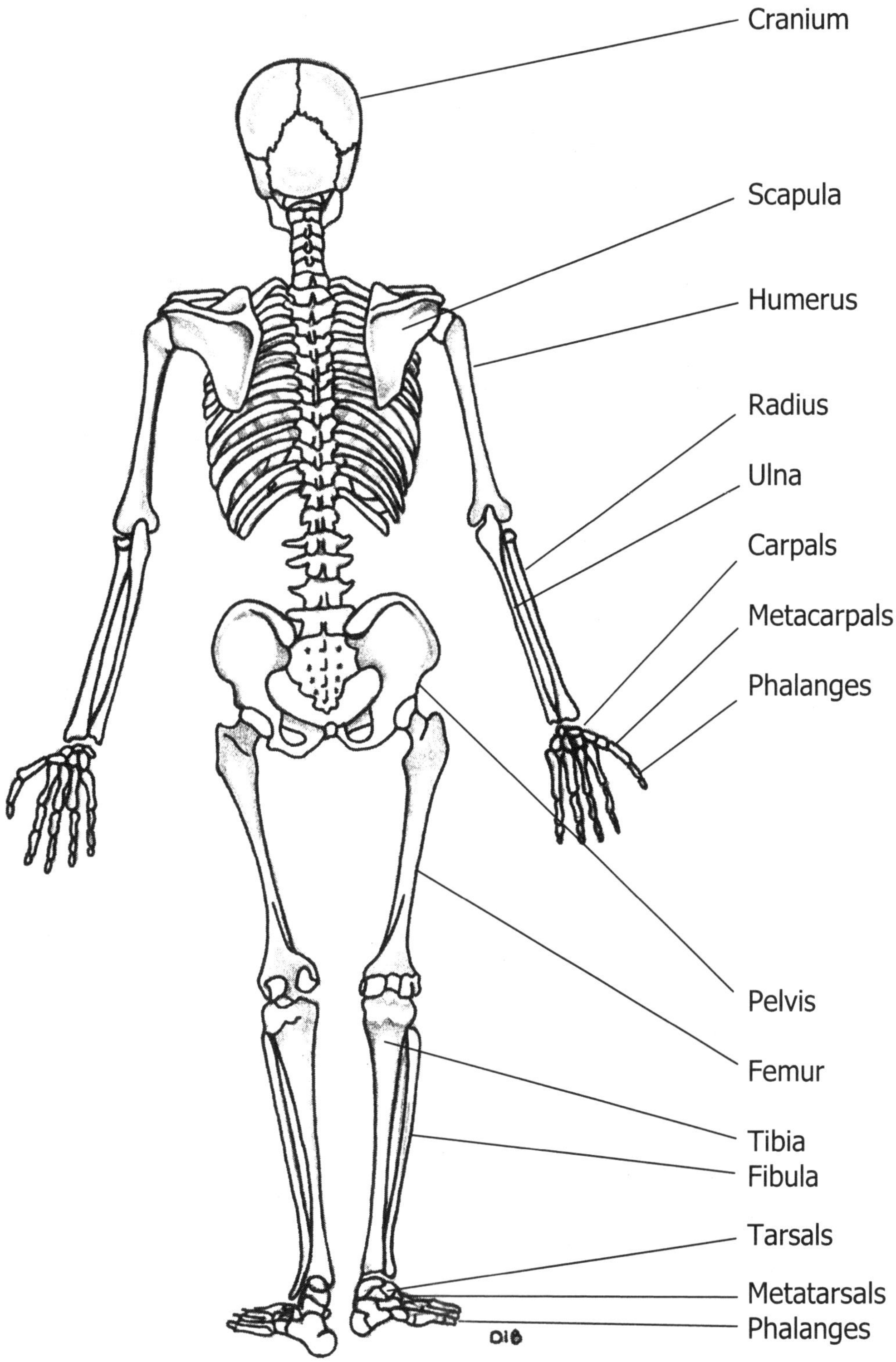

SKULL
Ventral (Anterior) View

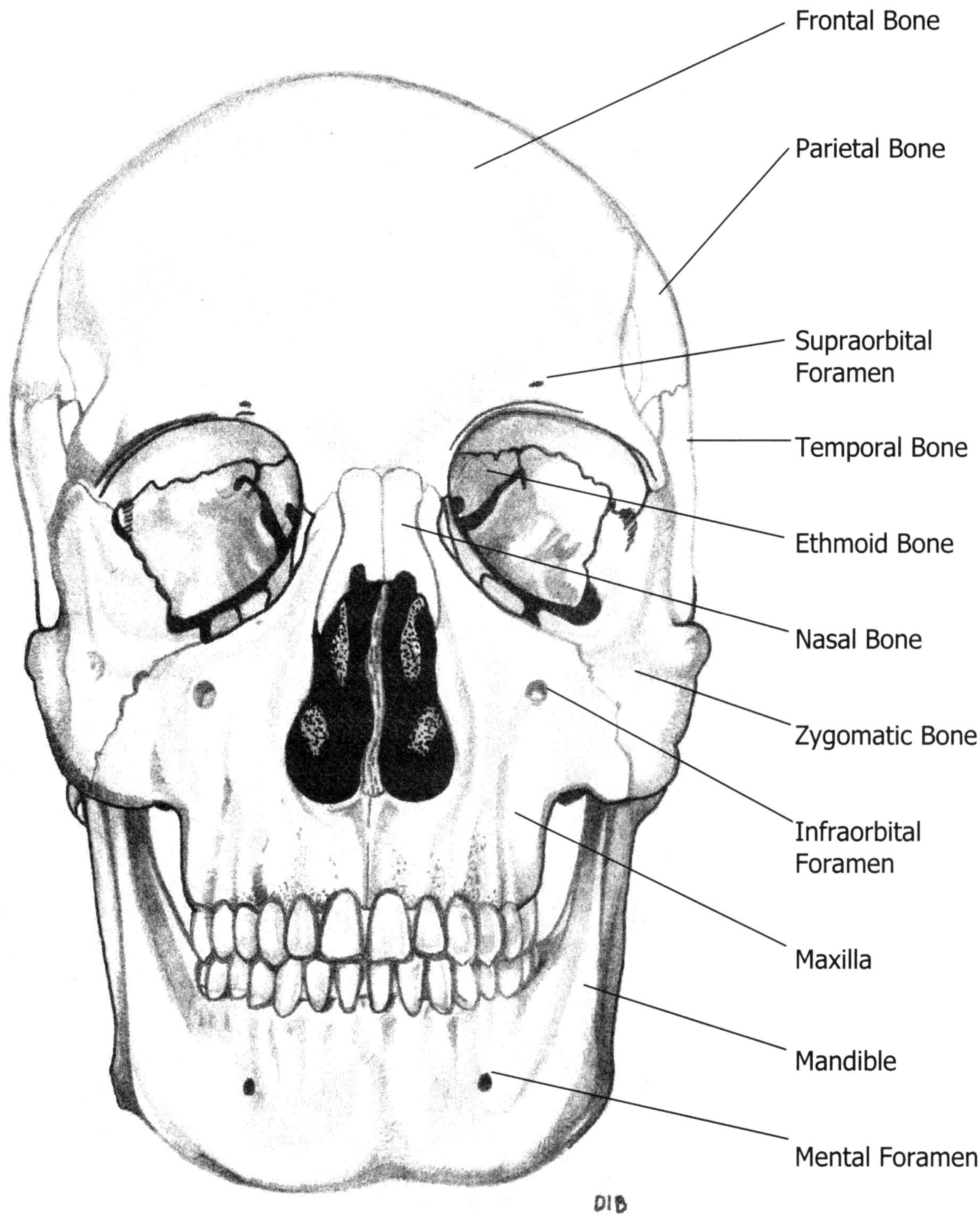

SKULL
(left) Lateral View

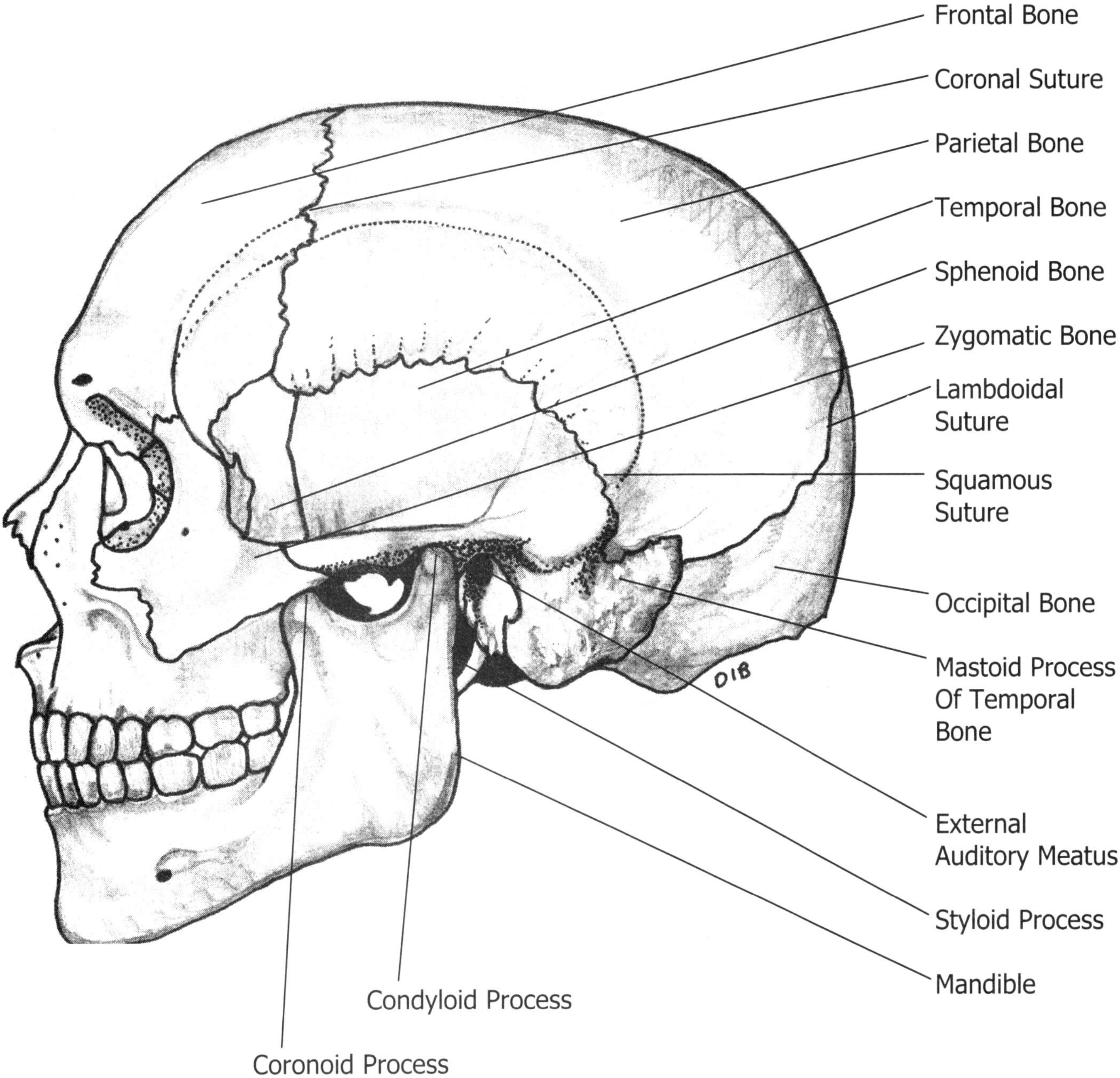

SKULL
Caudal (Inferior) View

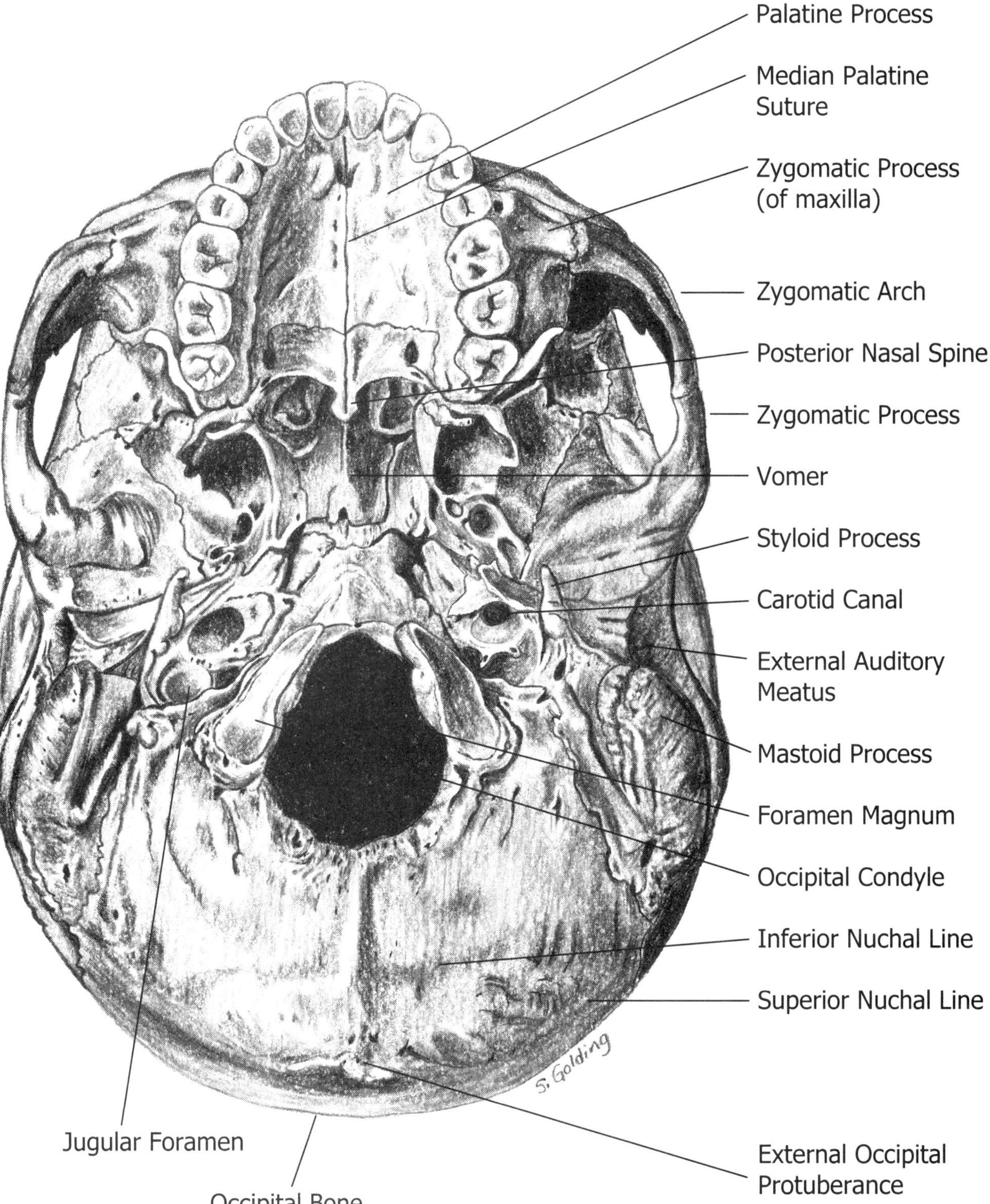

BONES of the MIDDLE EAR (Auditory Osicles)
Malleus, Incus, and Stapes

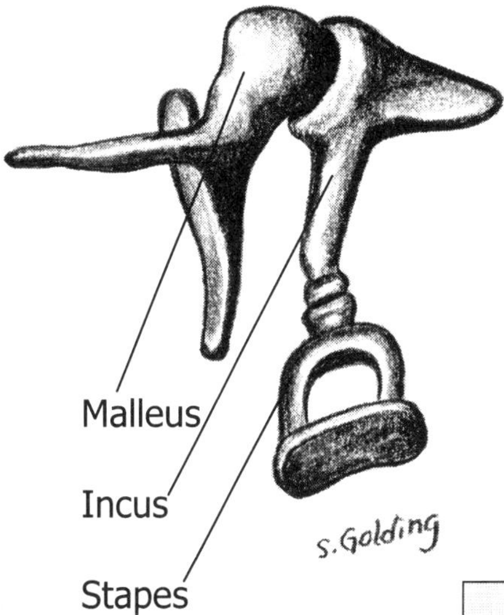

The Malleus, Incus and Stapes are so named for their shapes (ie: hammer, anvil, & stirrup) and are the smallest bones in the body. These tiny bones articulate via true synovial joints and function to transmit vibrations from the tympanic membrane (eardrum) to the oval window of the inner ear. Two of the smallest muscles in the body attach to the osicles and dampen loud noises.

HYOID BONE
Lateral and Anterior Views

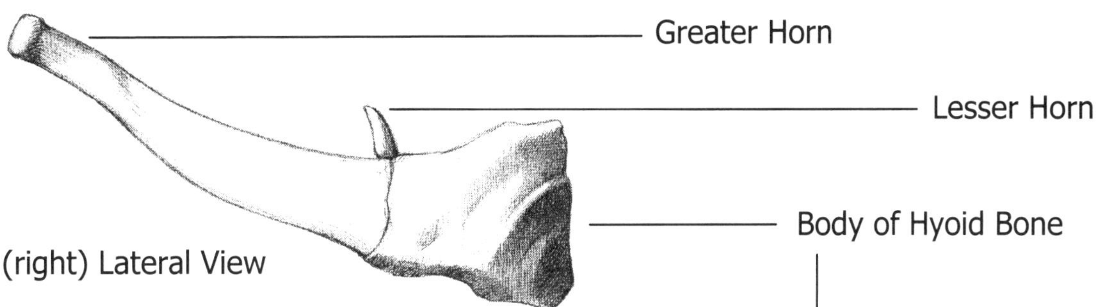

(right) Lateral View

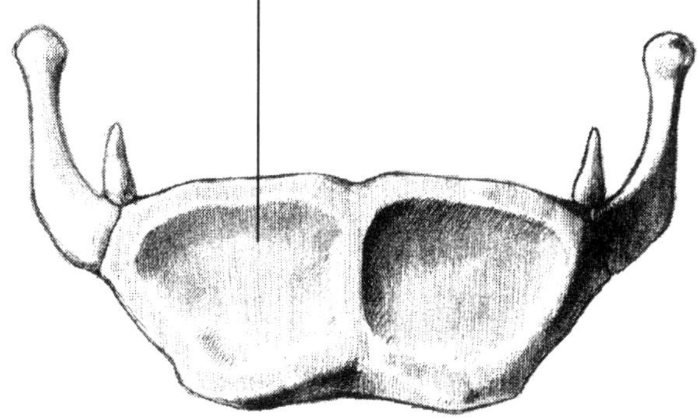

Anterior View

The single Hyoid bone (meaning U-shaped) is the only bone in the human body that does not articulate with another bone. It is suspended by ligaments and muscles from the styloid process of the temporal bone. The hyoid is found in the anterior neck, between the mandible and the larynx. It provides points of attachment for both the Suprahyoid and Infrahyoid muscles. The Hyoid bone is comprised of a horizontal body and two pairs of bony projections known as the greater and lesser horns. The horns provide for attachment of muscles and ligaments.

VERTEBRAL (SPINAL) COLUMN
Ventral (Anterior) View

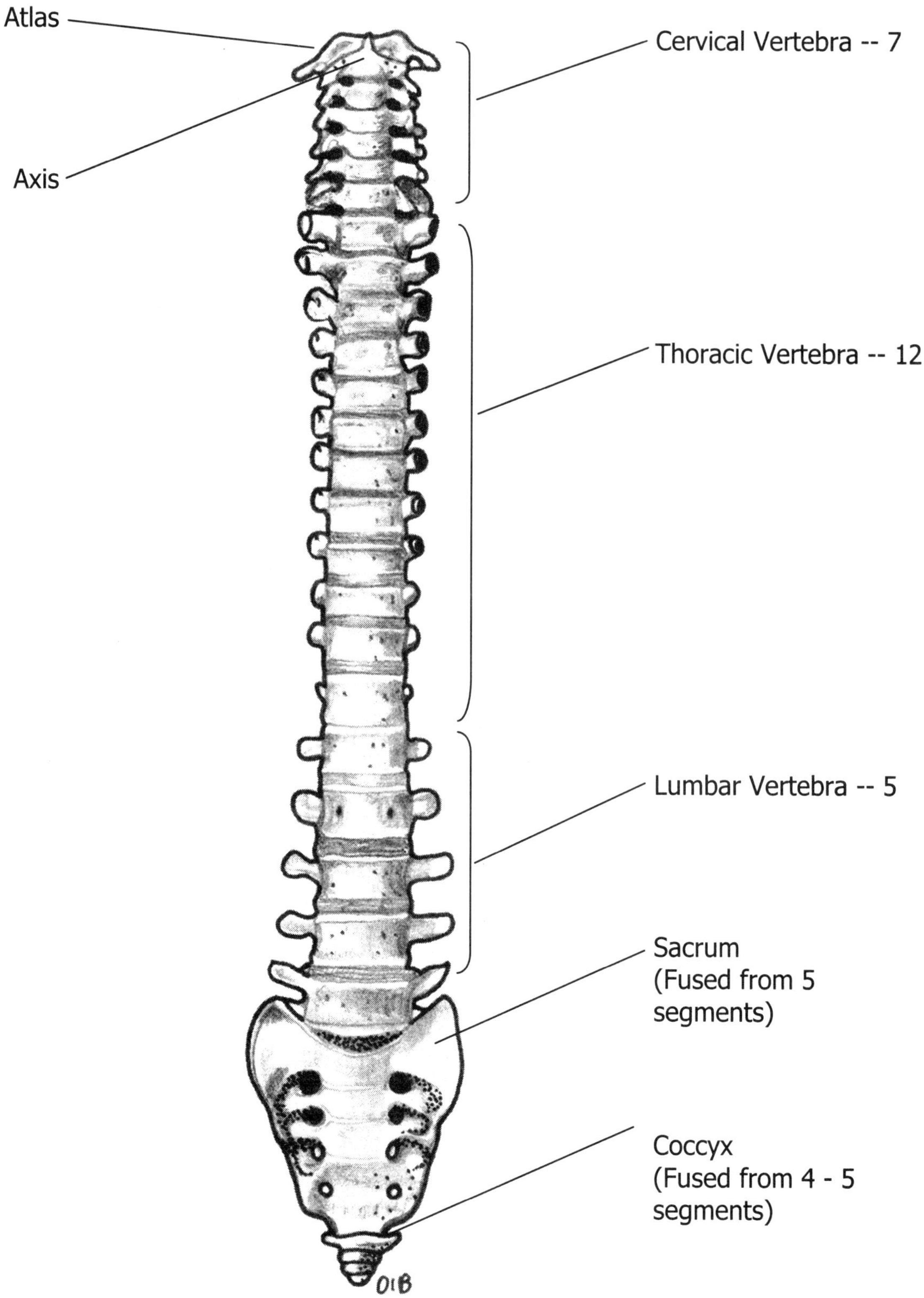

VERTEBRAL (SPINAL) COLUMN
(left) Lateral View

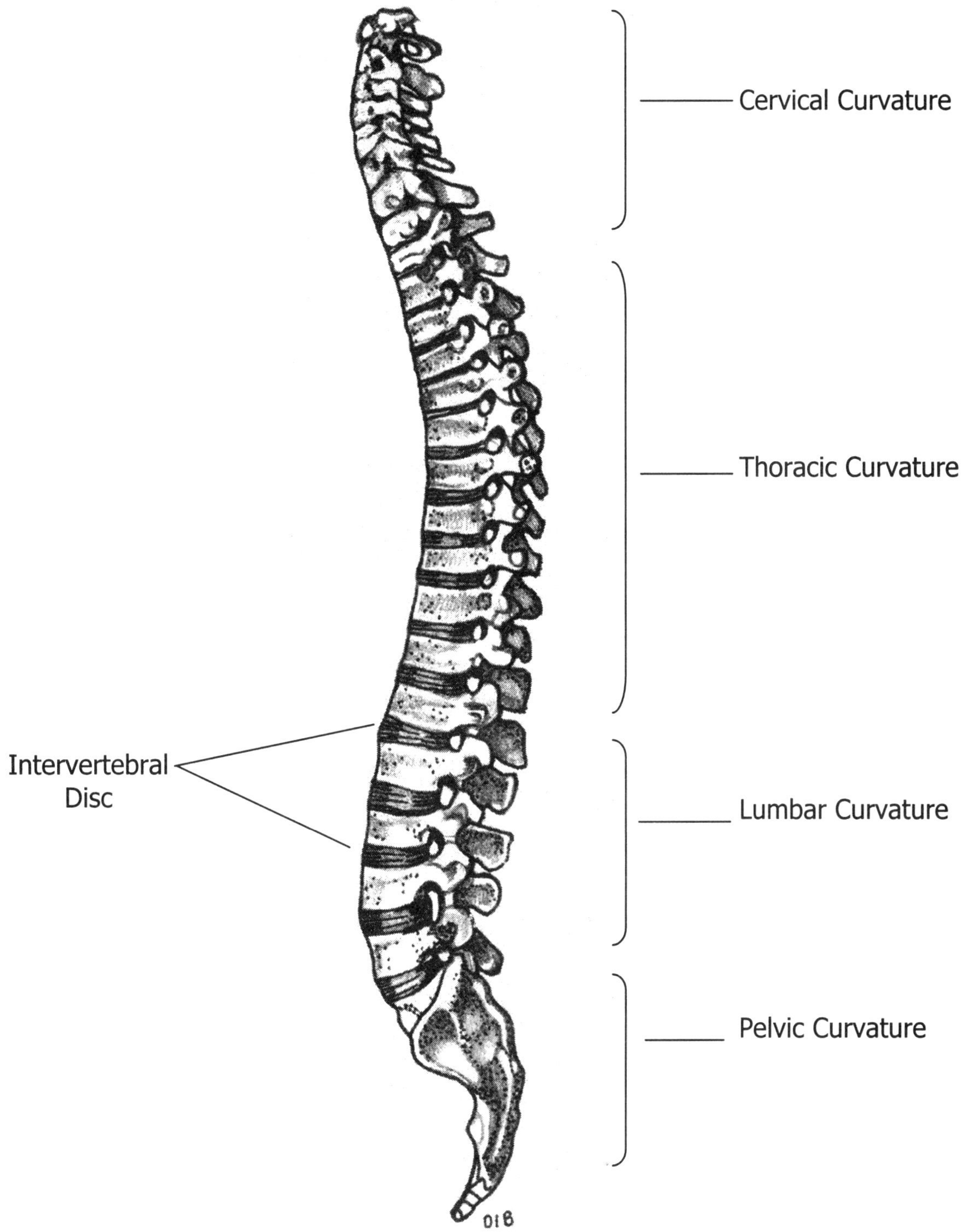

ATLAS (1st Cervical Vertebra)
Caudal (Inferior) View

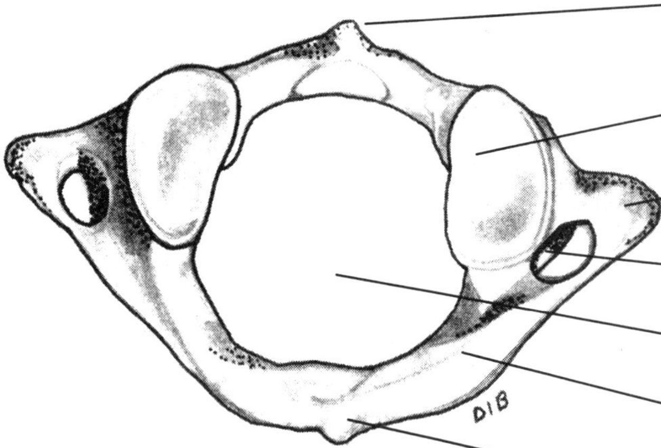

AXIS (2nd Cervical Vertebra)
(left) Lateral View

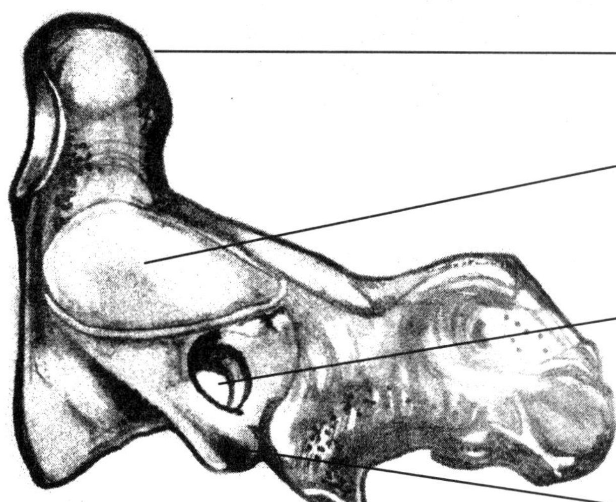

ATLAS & AXIS Articulated
(left) Sagittal Section

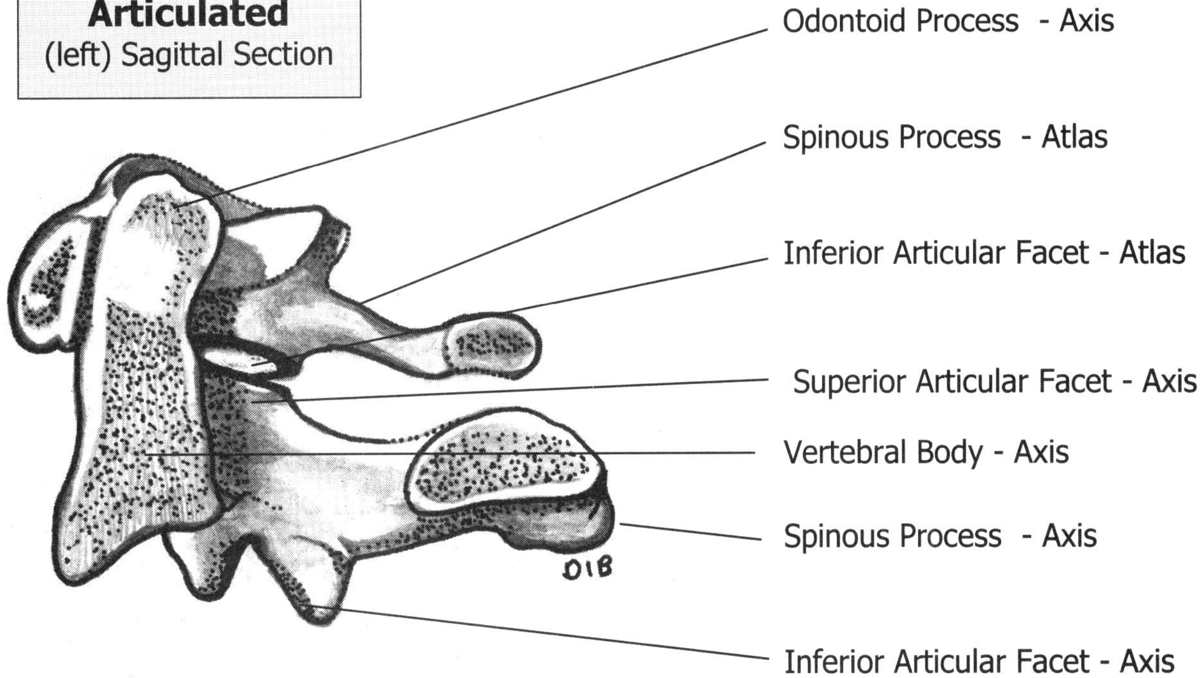

7th CERVICAL VERTEBRA (Vertebra Prominens)
Cephalic (Superior) View

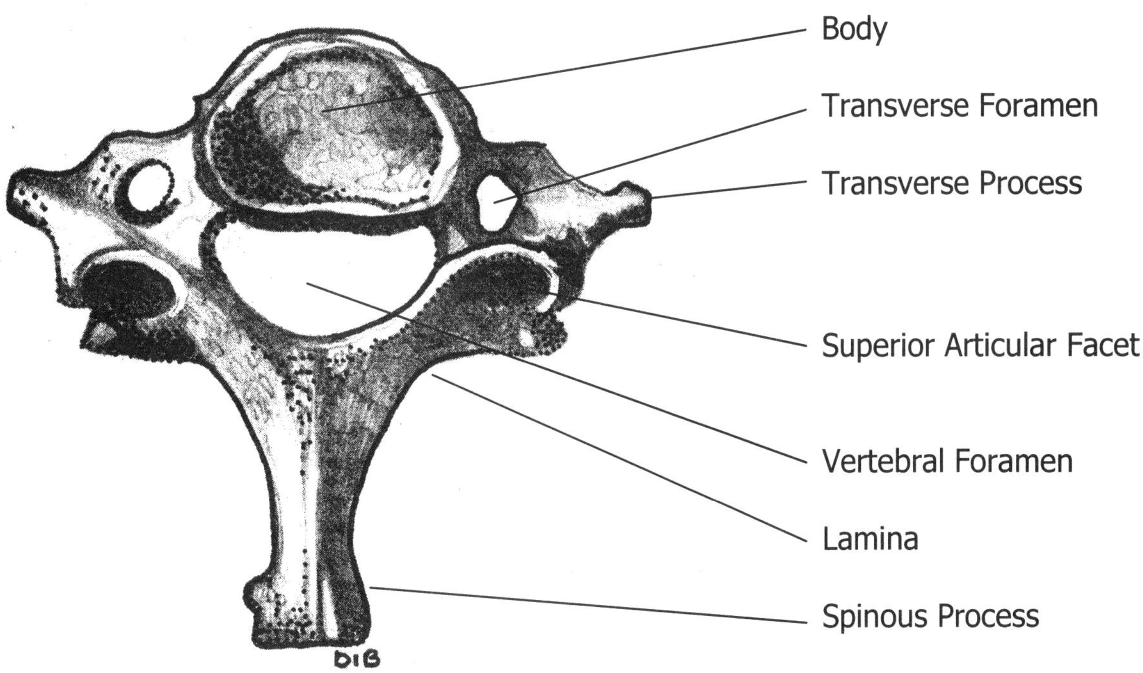

6th THORACIC VERTEBRA
Cephalic (Superior) View

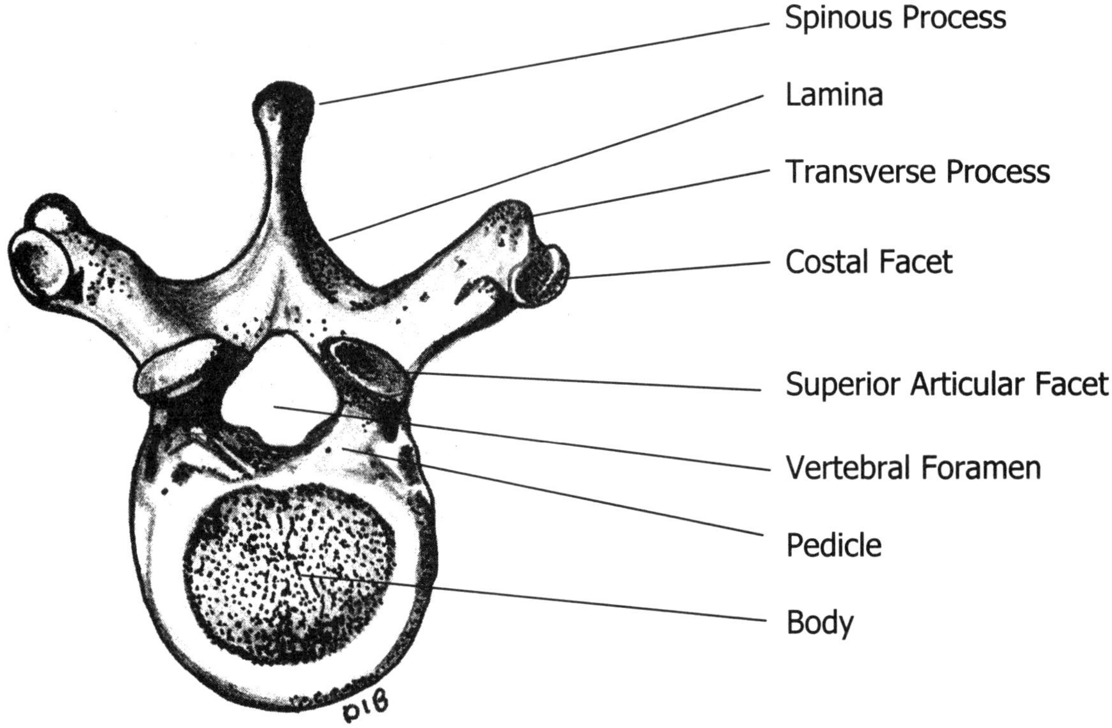

5th LUMBAR VERTEBRA
Cephalic (Superior) View

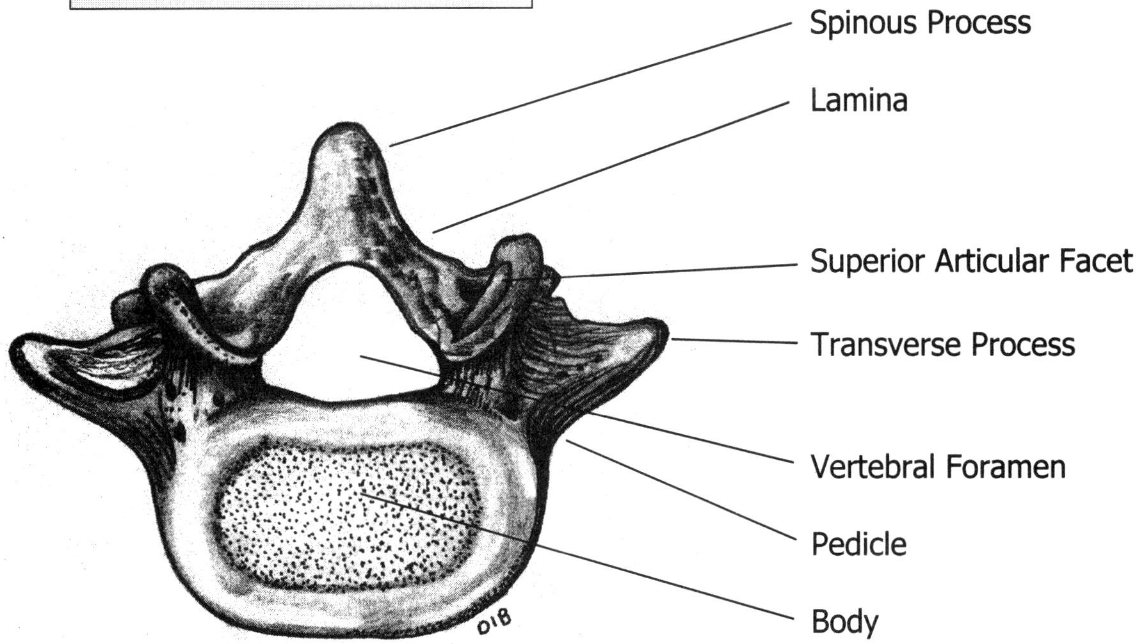

STERNUM & RIBS
Ventral (Anterior) View

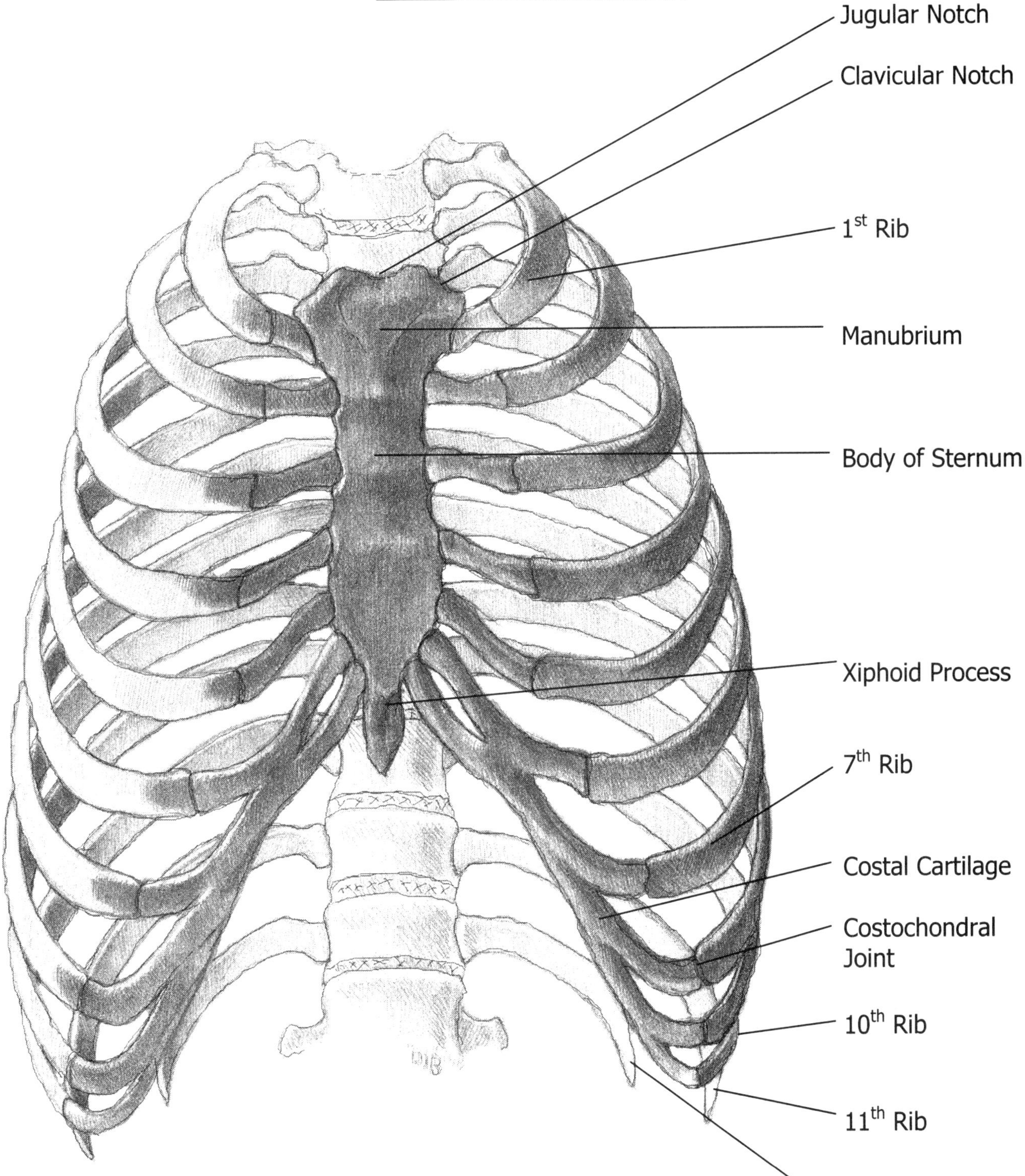

UPPER APPENDICULAR SKELETON (right)
Ventral (Anterior) View

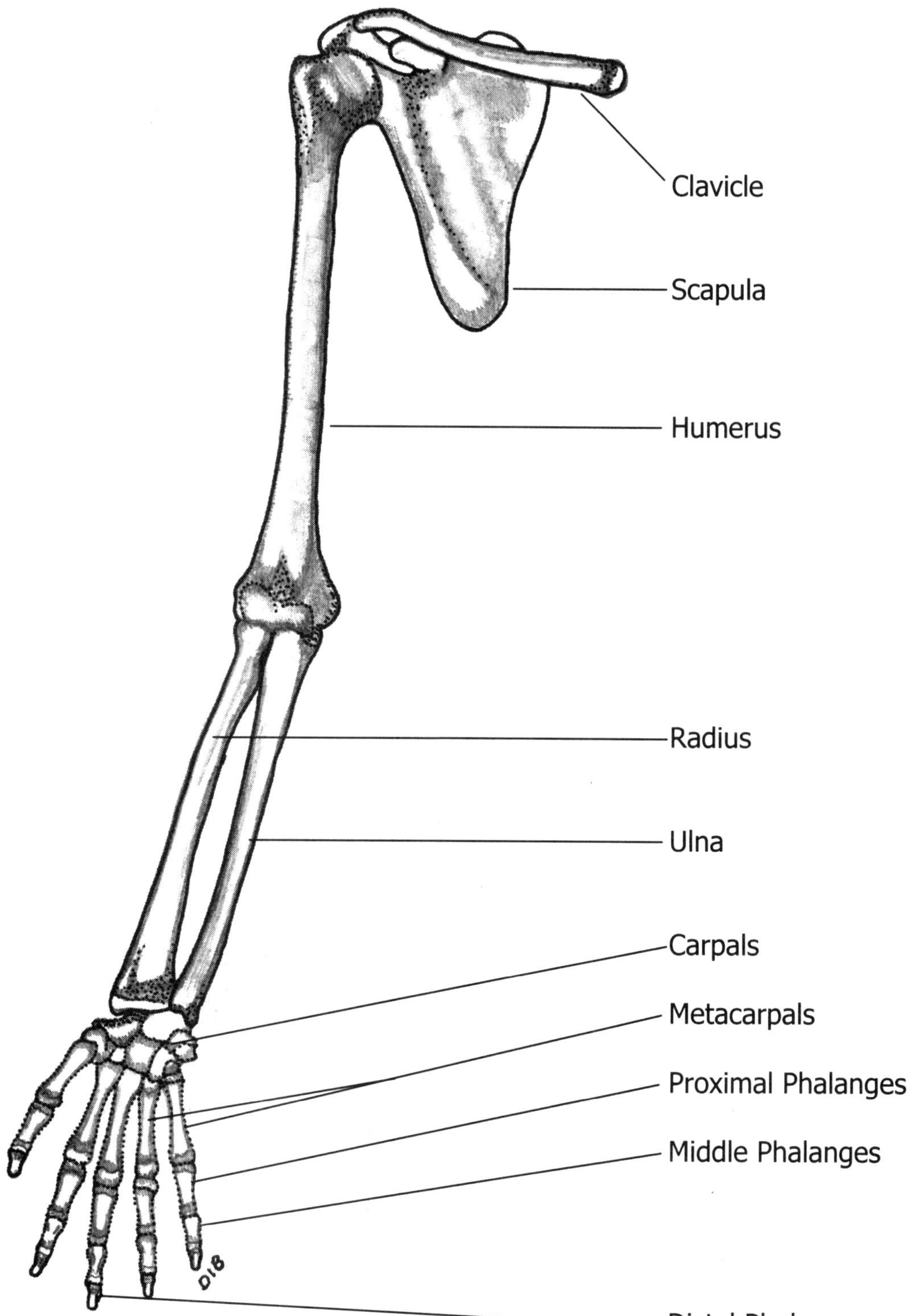

CLAVICLE (right)
Cephalic (Superior) View

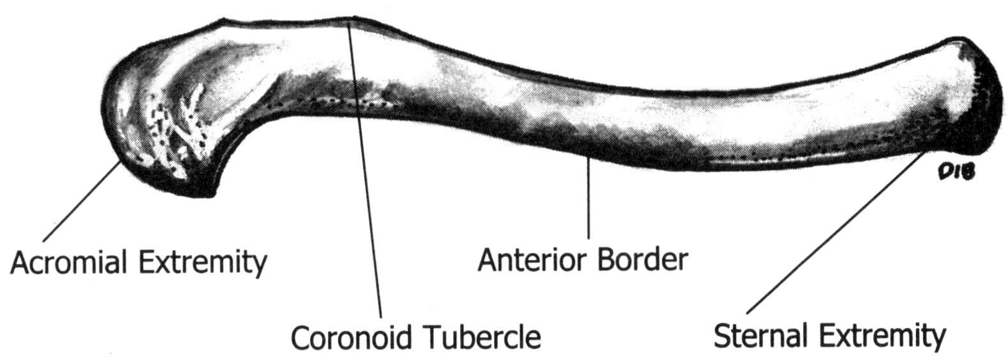

CLAVICLE (right)
Caudal (Inferior) View

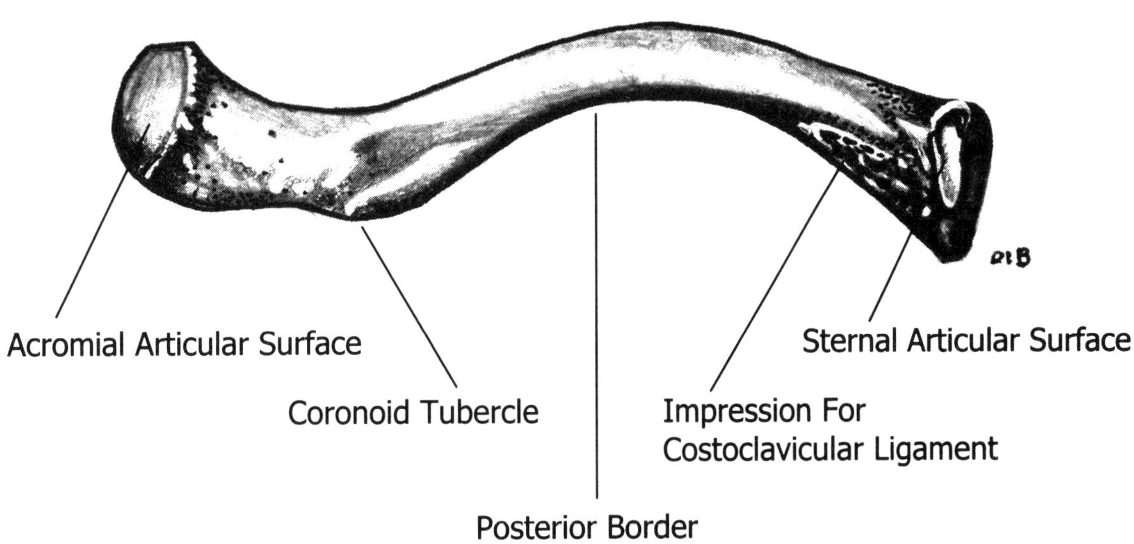

SCAPULA (right)
Dorsal (Posterior) View

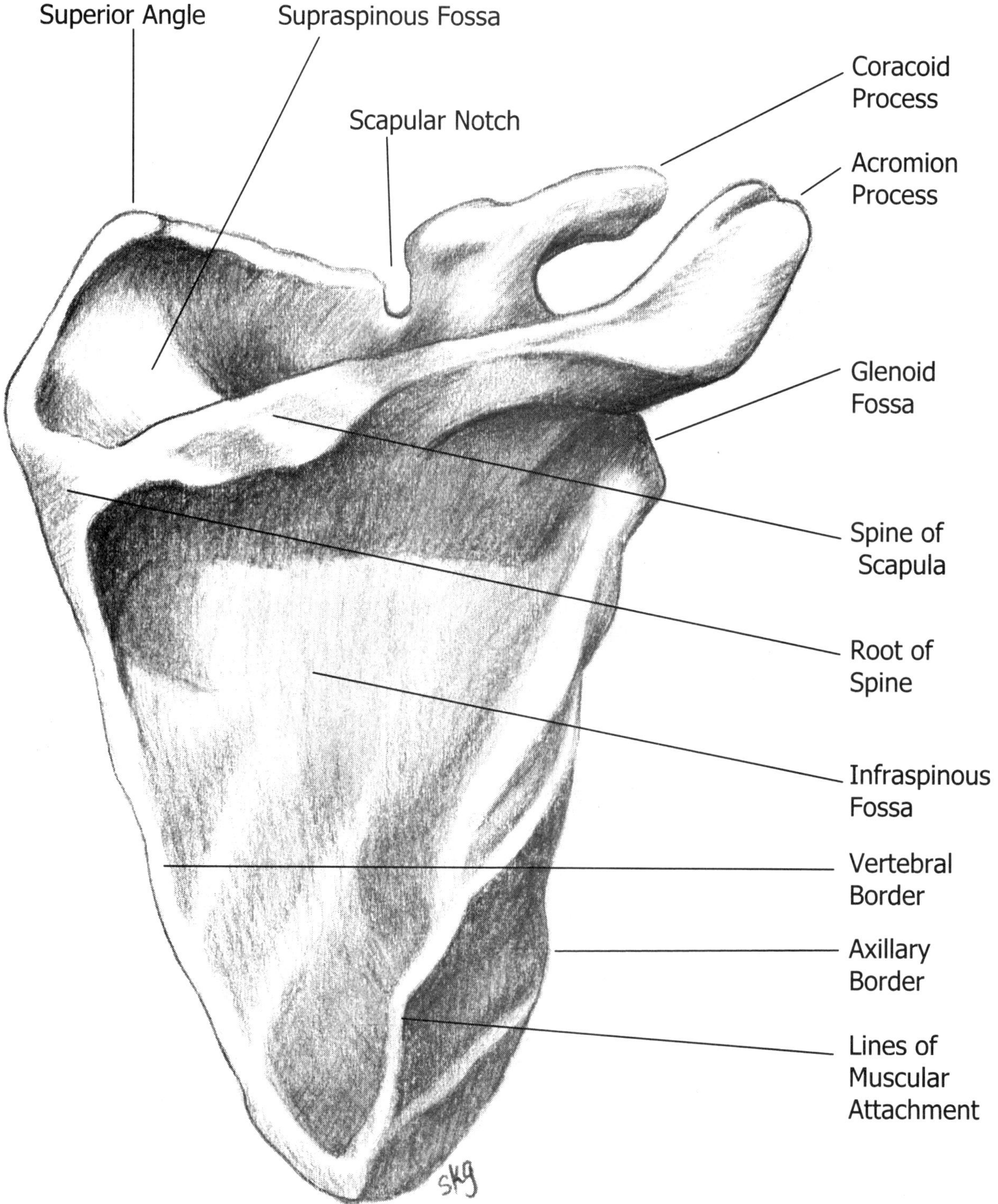

SCAPULA (right)
Ventral (Anterior) View

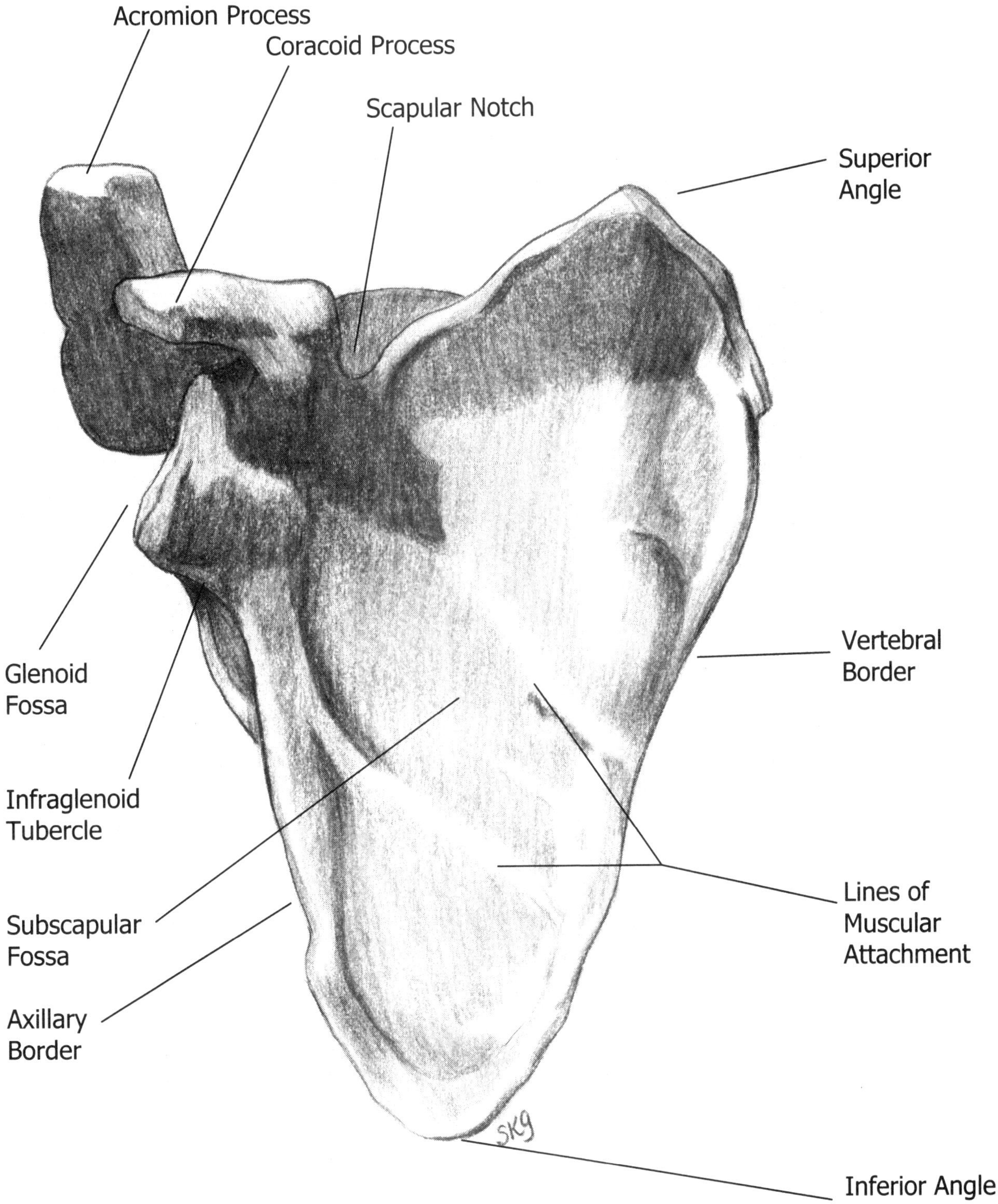

SCAPULA (right)
Medial to Lateral View

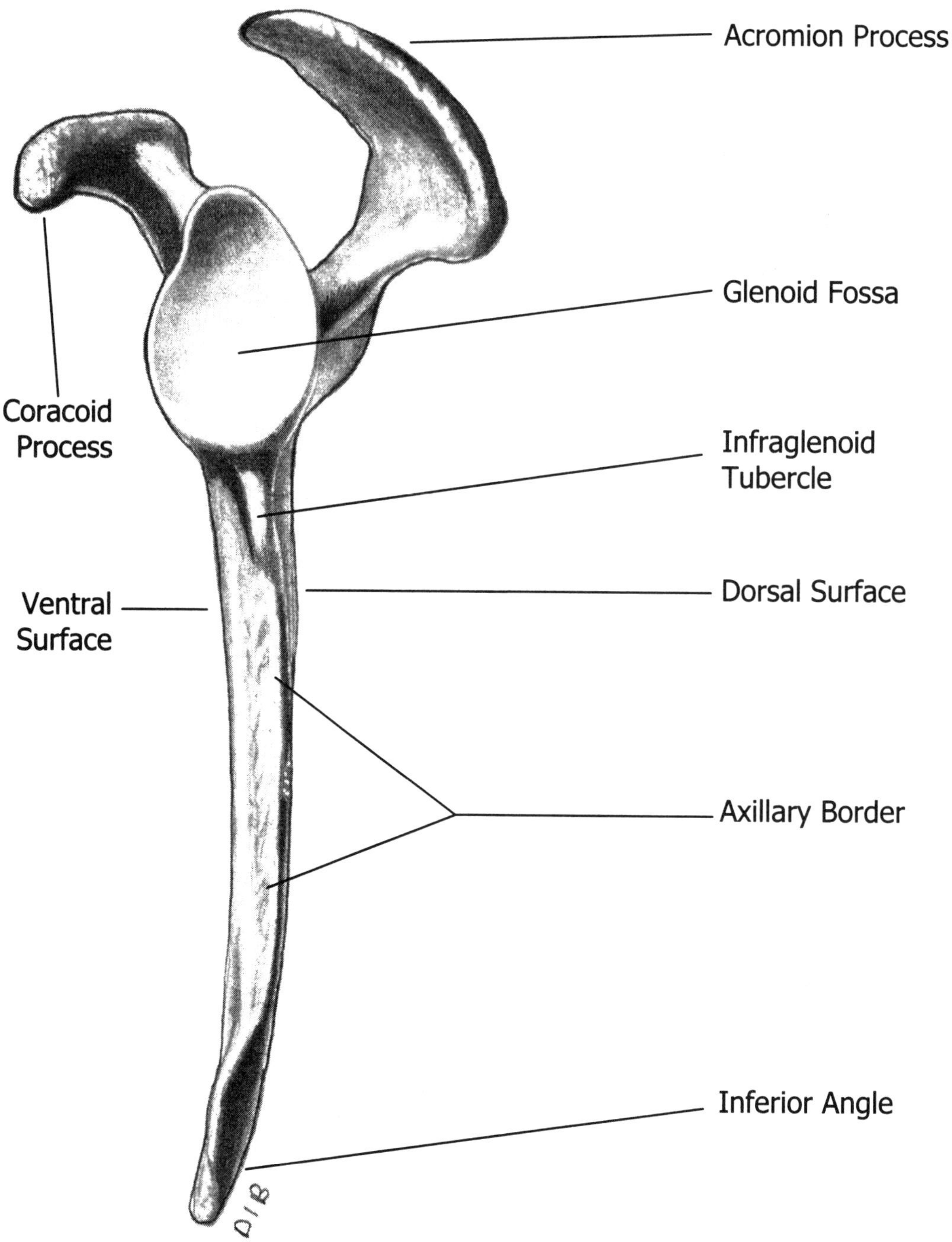

HUMERUS (left)
Ventral (Anterior) View

HUMERUS (left)
Dorsal (Posterior) View

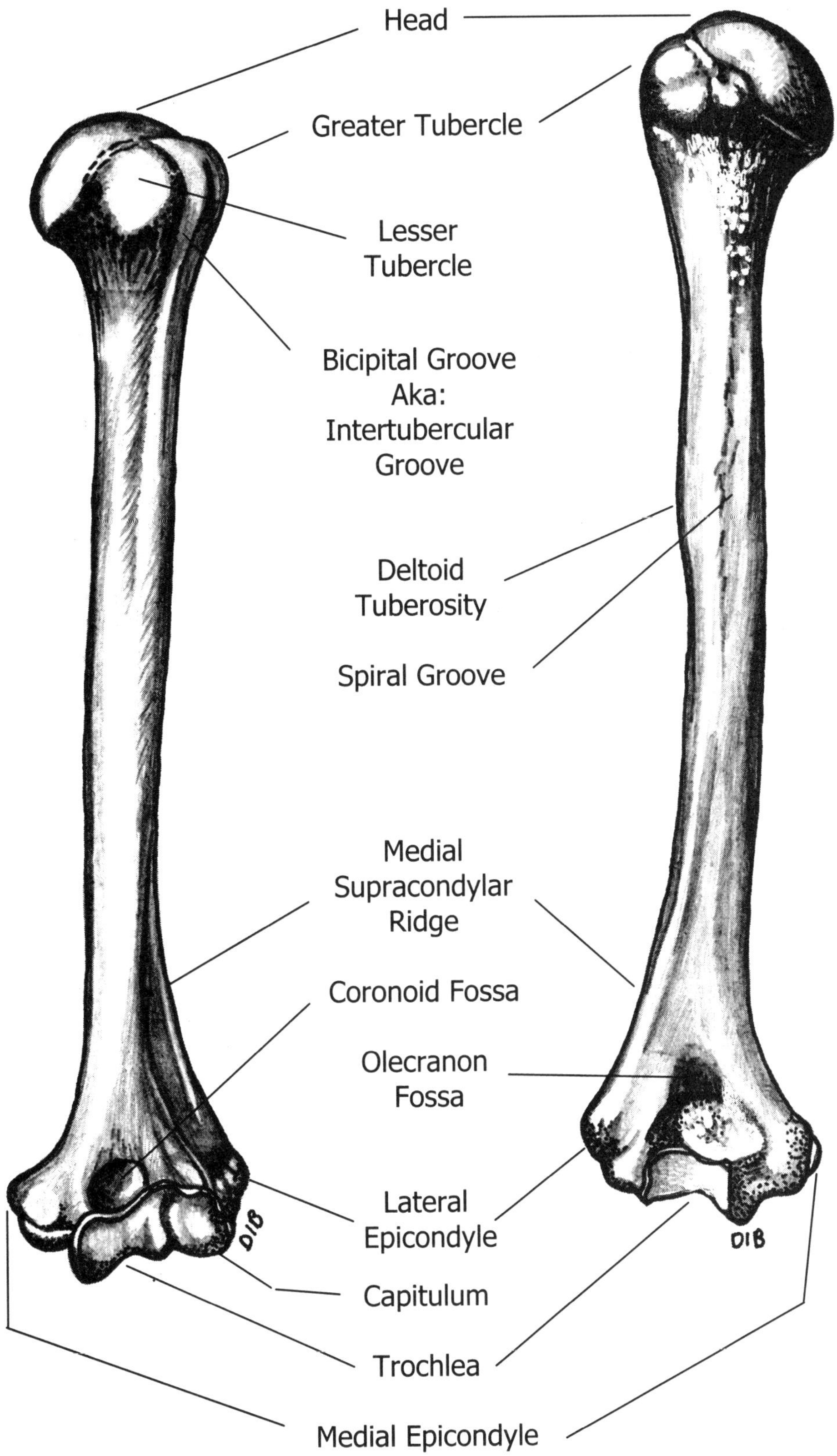

RADIUS (left)
Ventral (Anterior) View

ULNA (left)
Lateral View

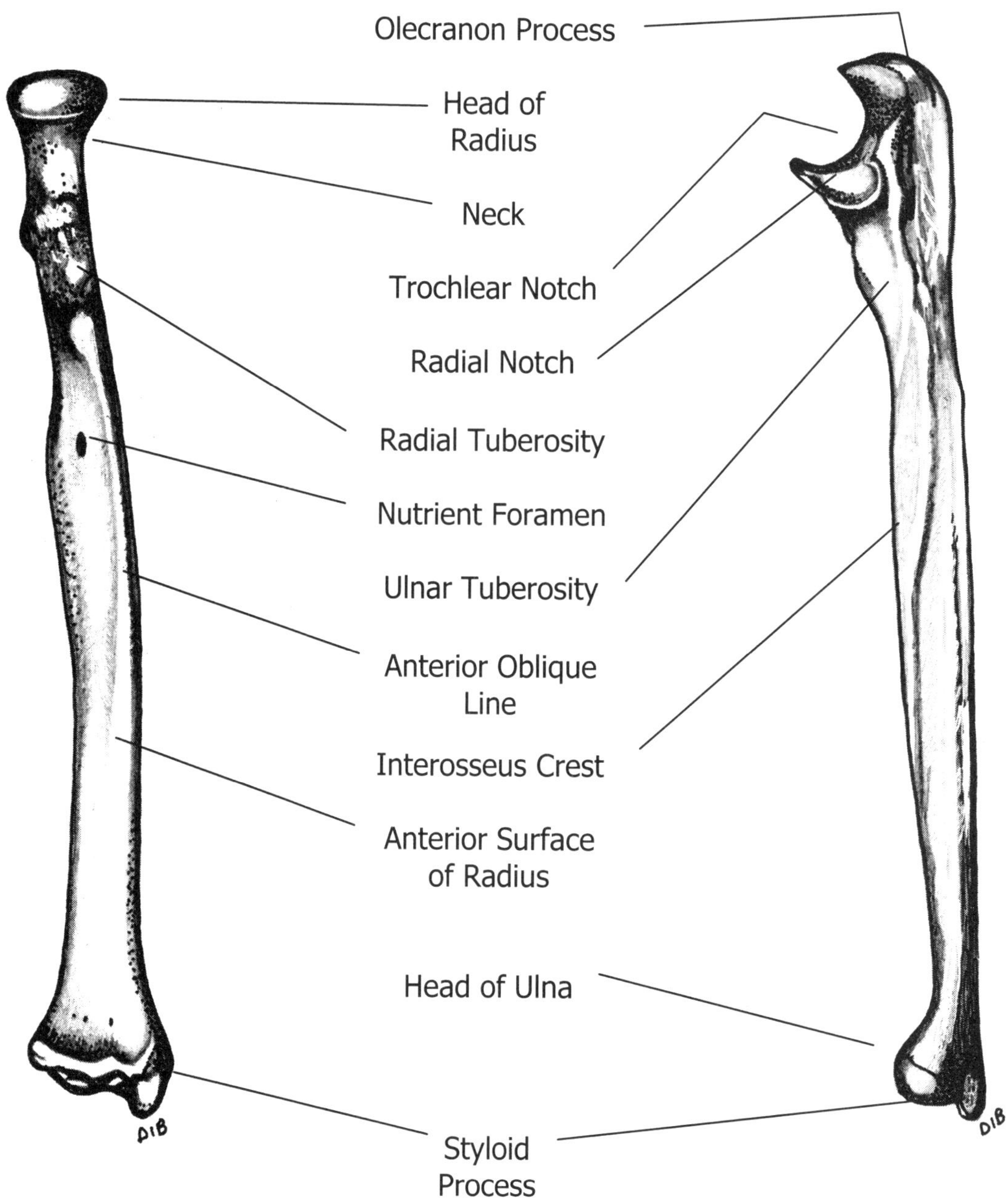

HAND (right) & CARPAL BONES
Palmar View

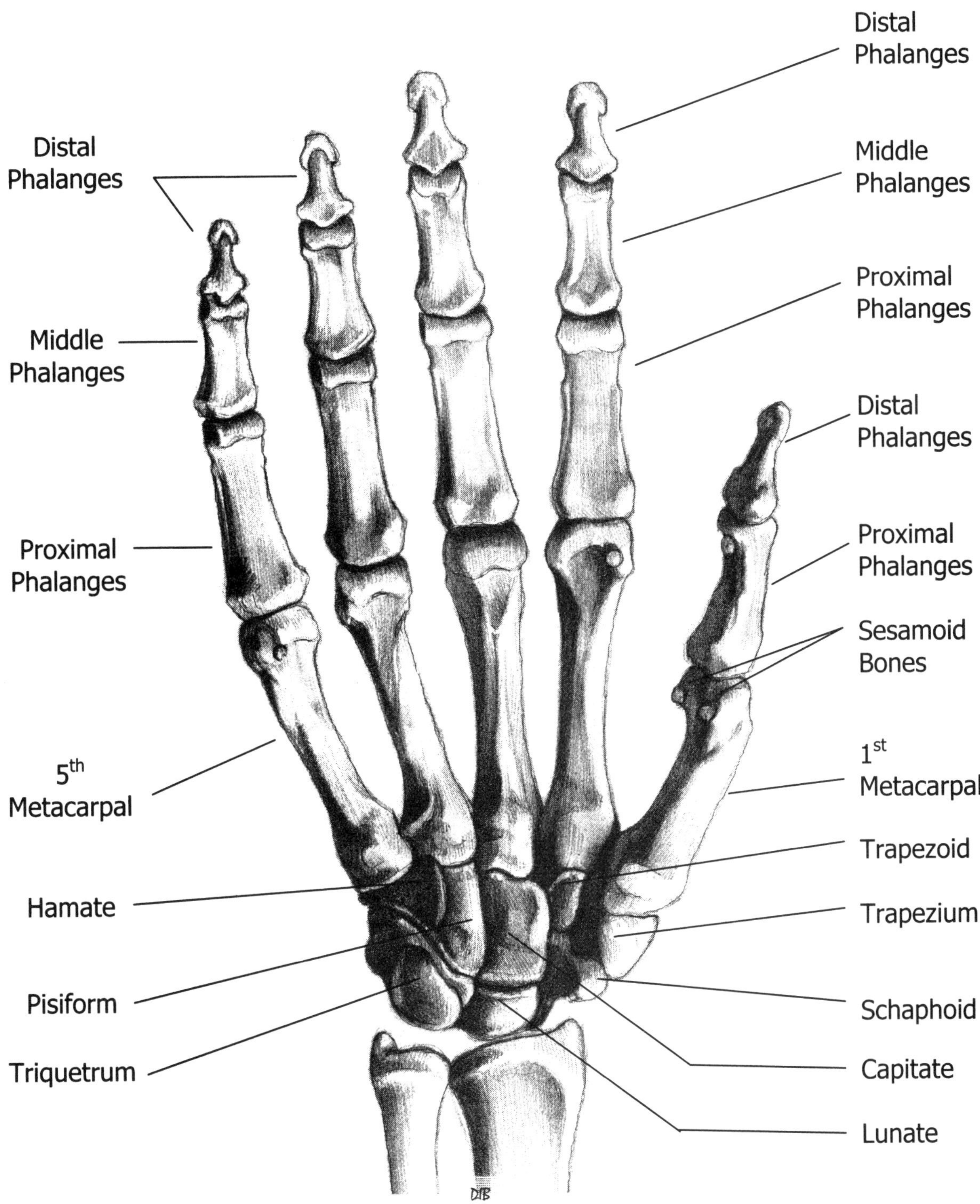

PELVIS (right) or **Innominate Bone**

Lateral View

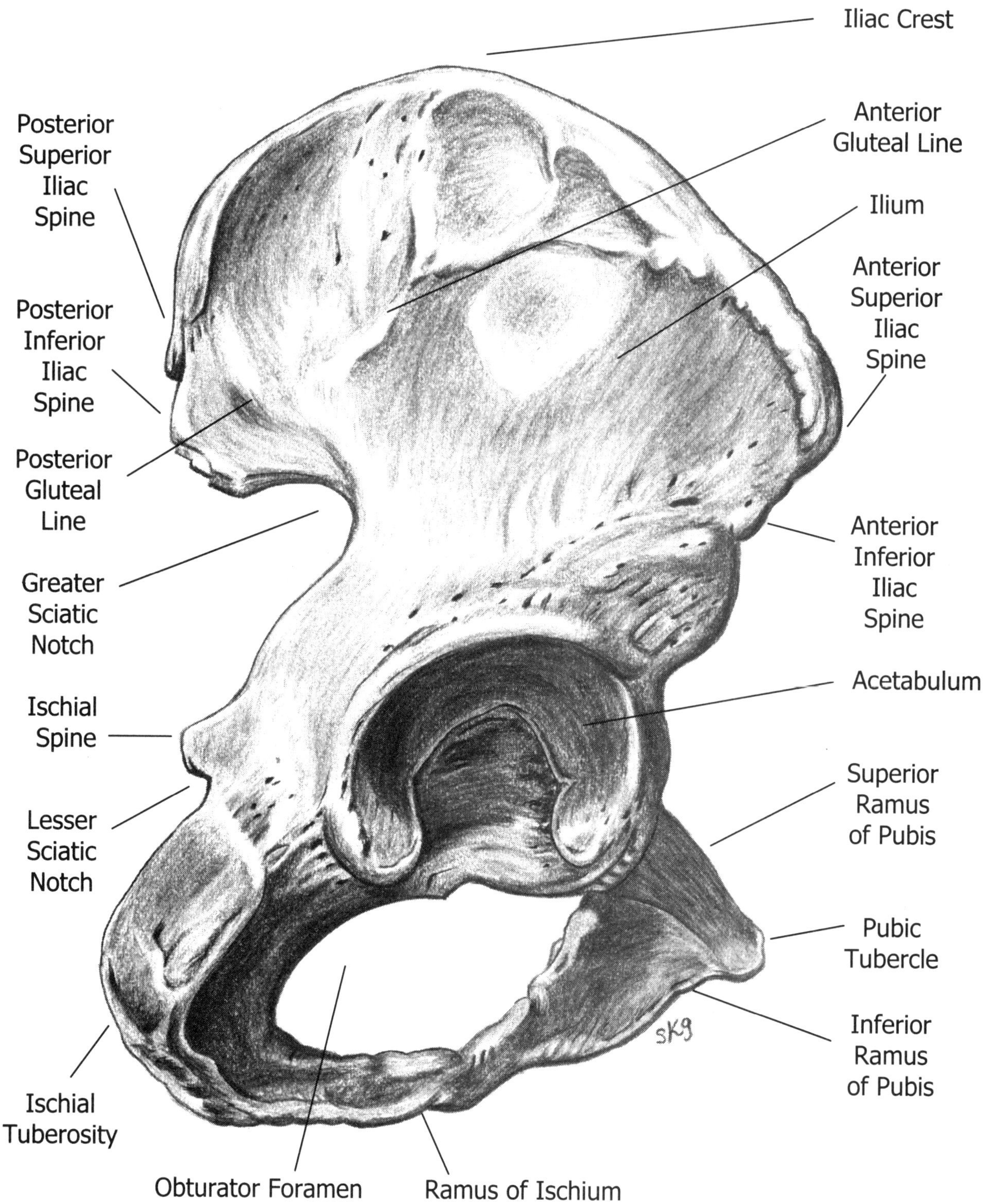

PELVIS (right) or
Innominate Bone
Medial View

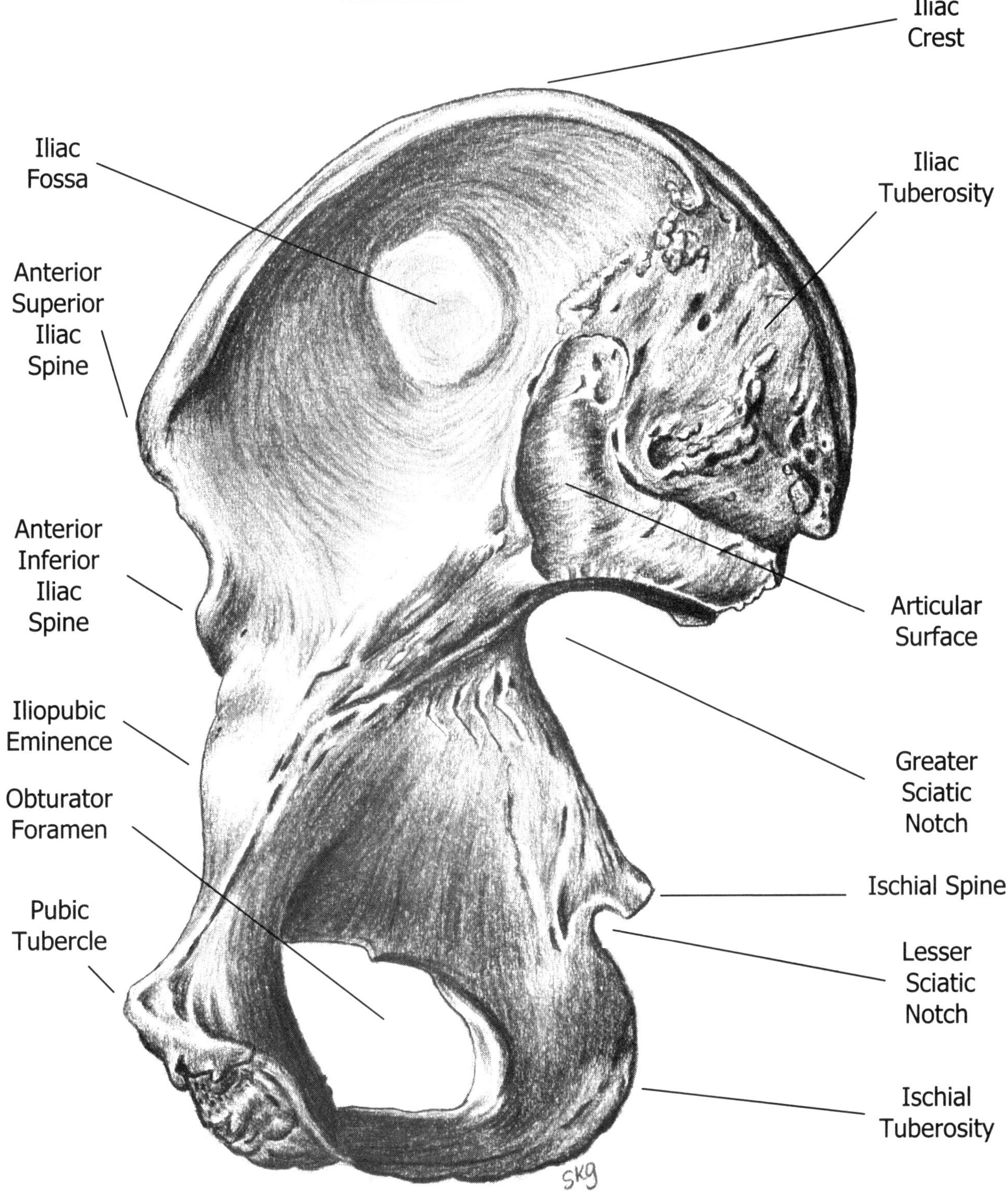

FEMUR (right)
Ventral (Anterior) View

FEMUR (right)
Dorsal (Posterior) View

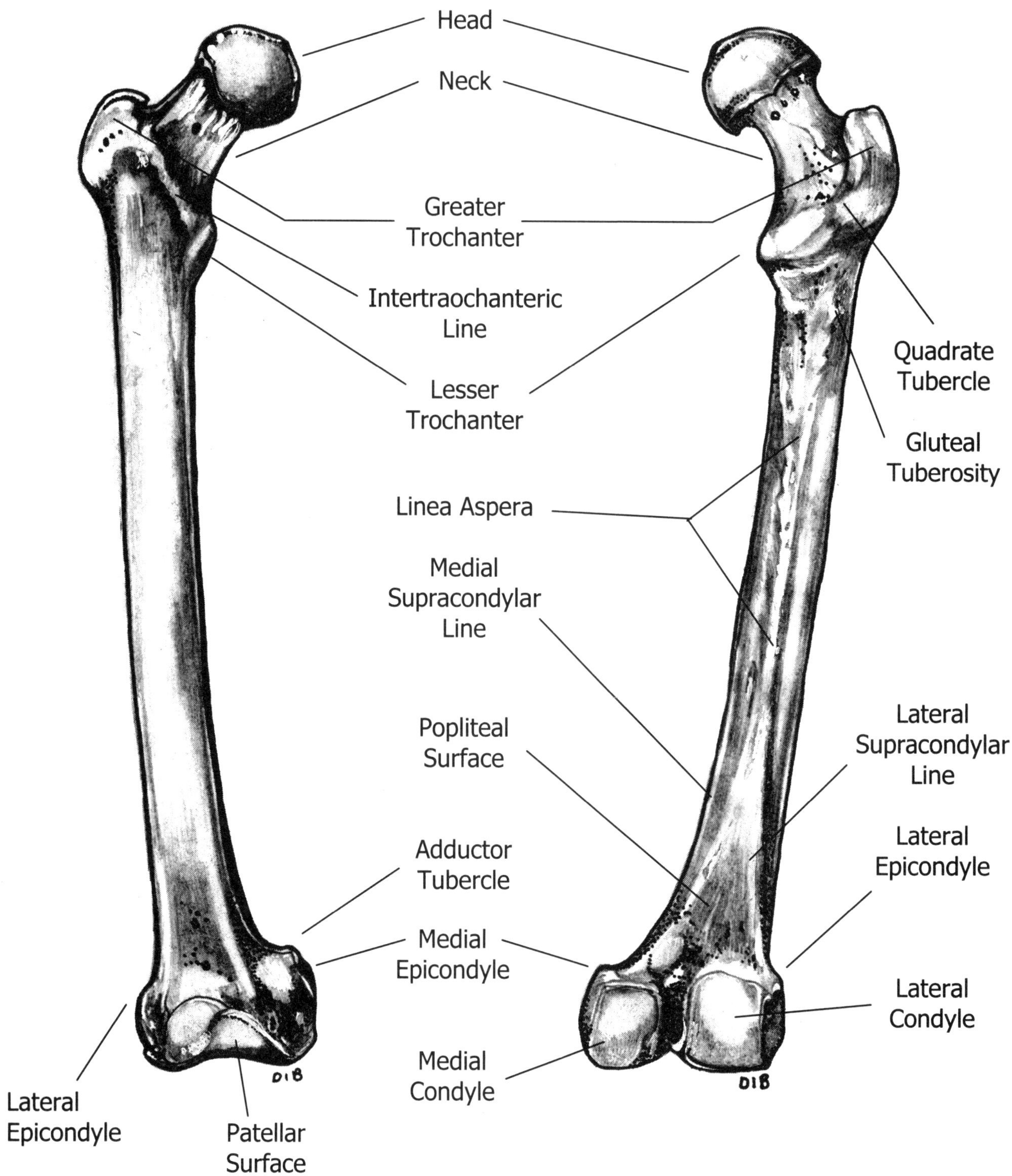

TIBIA & FIBULA (left)
Ventral (Anterior) View

TIBIA & FIBULA (left)
Dorsal (Posterior) View

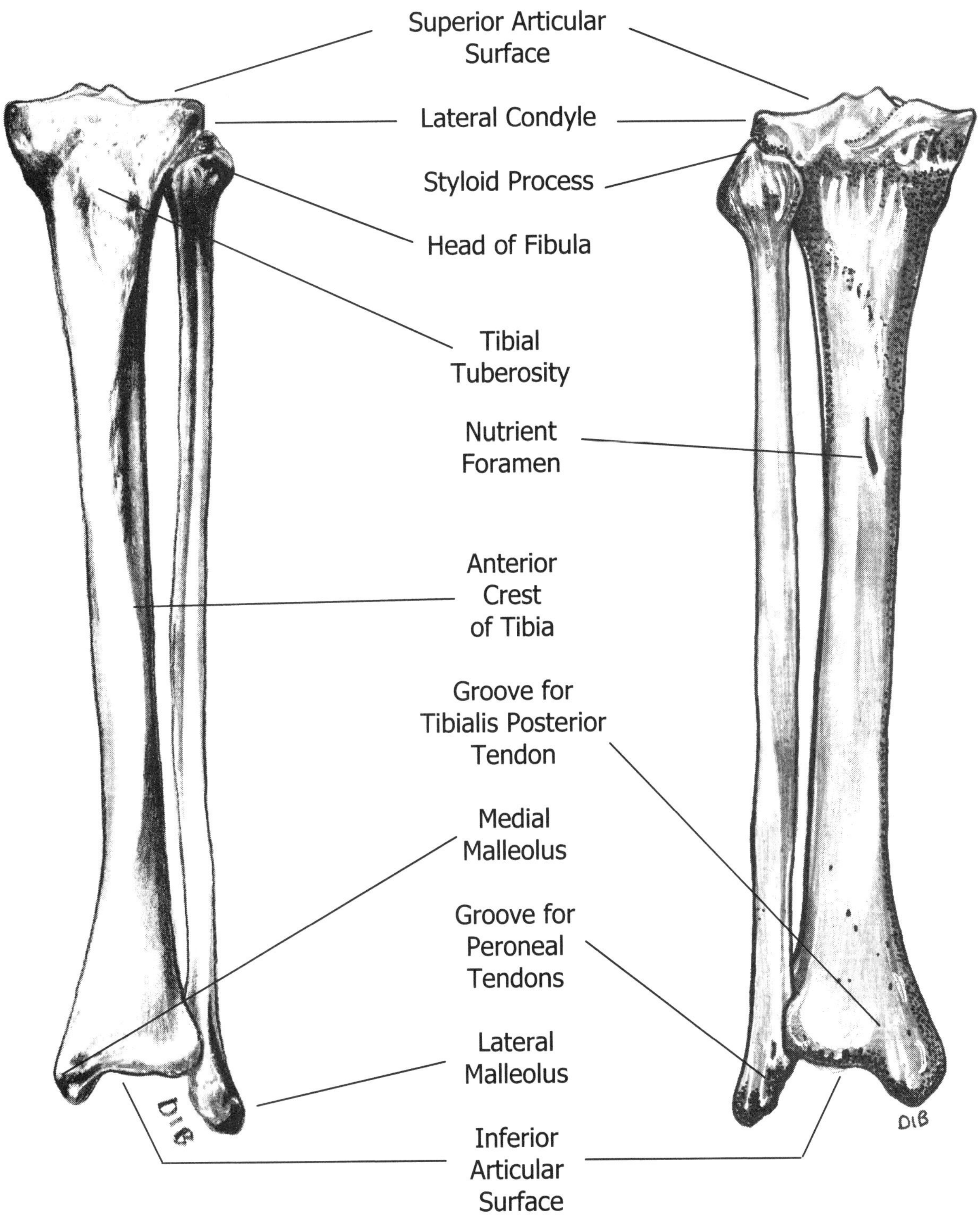

Foot (right)
Plantar View

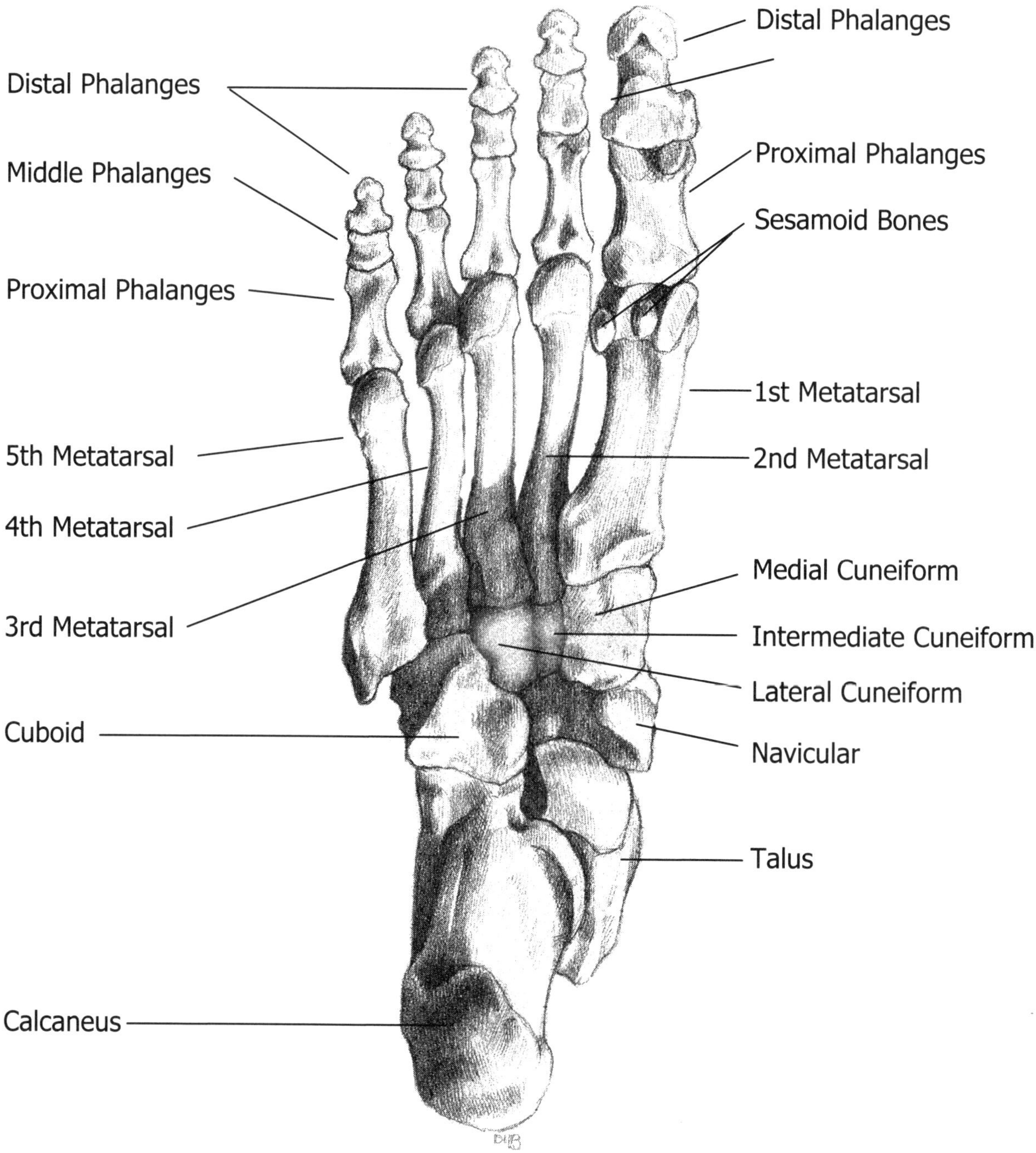

Foot (right)
Dorsal View

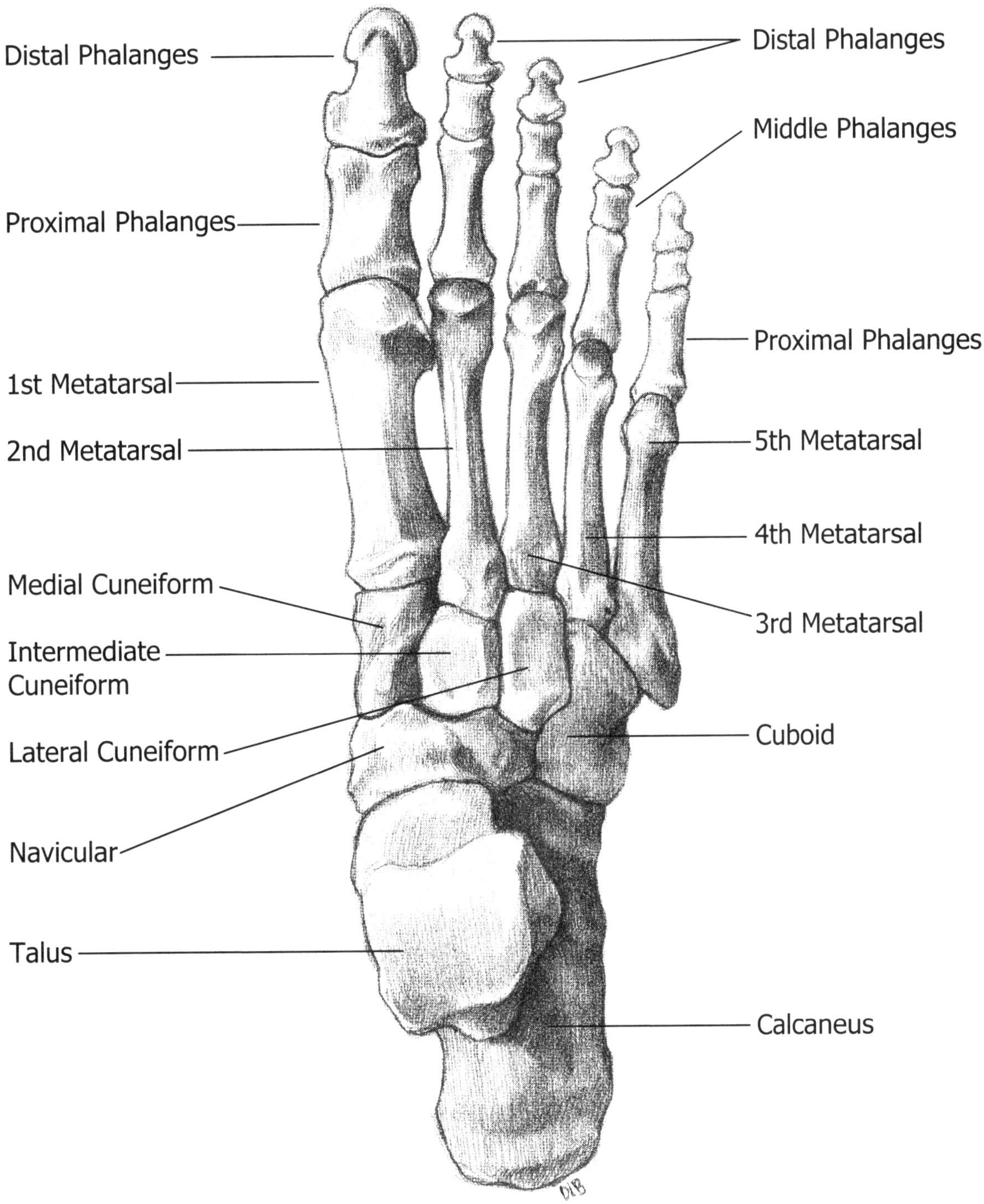

THE MUSCULAR SYSTEM

The human muscular system is composed of nearly 700 individual muscles, their tendons, and associated fascial structures. The following illustrations and descriptions are intended to familiarize you with the most important anatomical, biomechanical, and clinical attributes of each of the major muscles presented.

With each of the muscle illustrations presented, you will find described the following details:

"**Origin**" -- more stable and often more proximal of the muscle's attachments.

"**Insertion**" -- more movable of the muscle's attachments, with the insertion usually moving toward the origin as the muscle contracts.

"**Action**" -- the principal movement (or performance) resulting from that muscle's contraction. Some muscles cross only one joint and are responsible for a relatively simple movement, while other muscles cross two or more joints, resulting in complex movements.

"**Innervation**" -- motor nerve or branch that supplies the specific muscle.

"**Relationship**" -- in this case, meaning a description of the muscle shape, direction of fibers, location relative to other structures, activities involving the use of the muscle, major clinical signs & symptoms of muscular pain & dysfunction, and major differential diagnosis.

A *trigger point* as described herein has been defined by Drs. Travell and Simons as "a hyperirritable spot in skeletal muscle that is associated with a hypersensitive palpable nodule in a taut band. The spot is painful upon compression and can give rise to characteristic referred pain, referred tenderness, motor dysfunction, and autonomic phenomenon."

TRAPEZIUS (left)
Posterior View

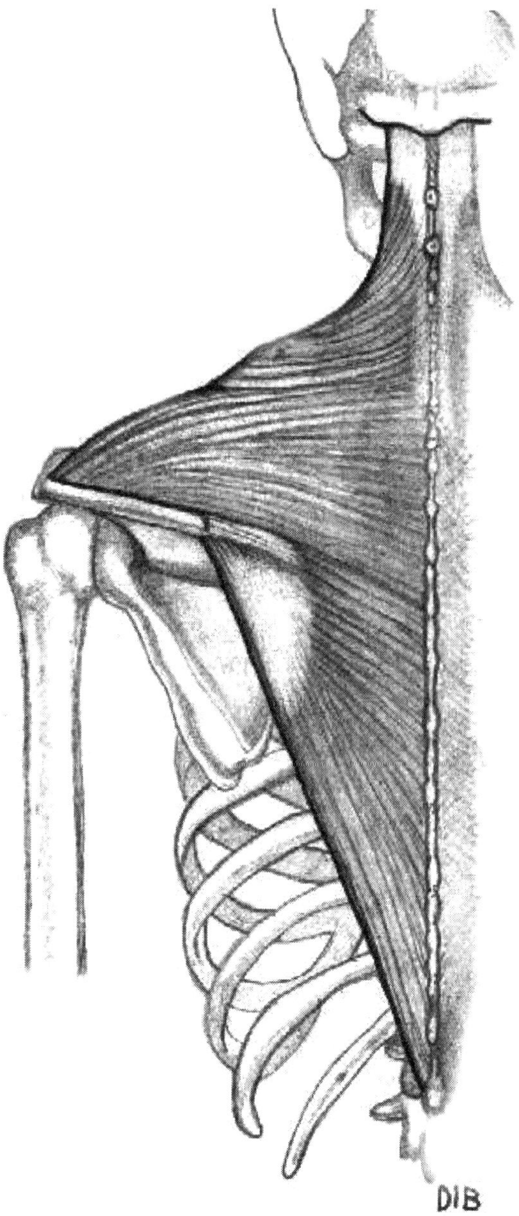

Origin:
Upper: Medial third of superior nuchal line, E.O.P. & ligamentum nuchae
Middle: C-6 (or C-7) through T-3 (spinous processes & ligaments)
Lower: T-4 through T-12 (spinous processes)

Insertion:
Upper: Posterior border of lateral third of clavicle
Middle: Medial margin of acromion & superior lip of spine of the scapula
Lower: Medial portion of spine of the scapula

Action:
Upper: Extends & laterally flexes head and neck to same side, aids in rotation to opposite side, elevates the sternoclavicular joint & glenoid fossa
Middle: Retraction of scapula and elevation of glenoid fossa
Lower: Stabilization of scapula, aiding in elevation of the glenoid fossa

Innervation:
Spinal accessory nerve (cranial nerve # XI) & branches of spinal nerves C-2, C-3 & C-4

Relationship:
Trapezius is a flat, triangular, superficial muscle of the upper back, with the paired trapezius forming a diamond shape. It functions as three distinct muscles, with fibers that travel in three directions. Trapezius participates in most activities requiring movements of the arm and steadying of the scapula. Rowing exercises and shoulder shrugs strengthen this muscle. Upper trapezius is often involved in whiplash type injuries and presents with constant neck pain and headache. Middle and lower trapezius involvement commonly present with burning interscapular pain. Symptoms must be differentiated from those of tension headache, cervical arthralgia, TMJ disorder, and fibromyalgia.

LATISSIMUS DORSI
(left) Posterolateral View

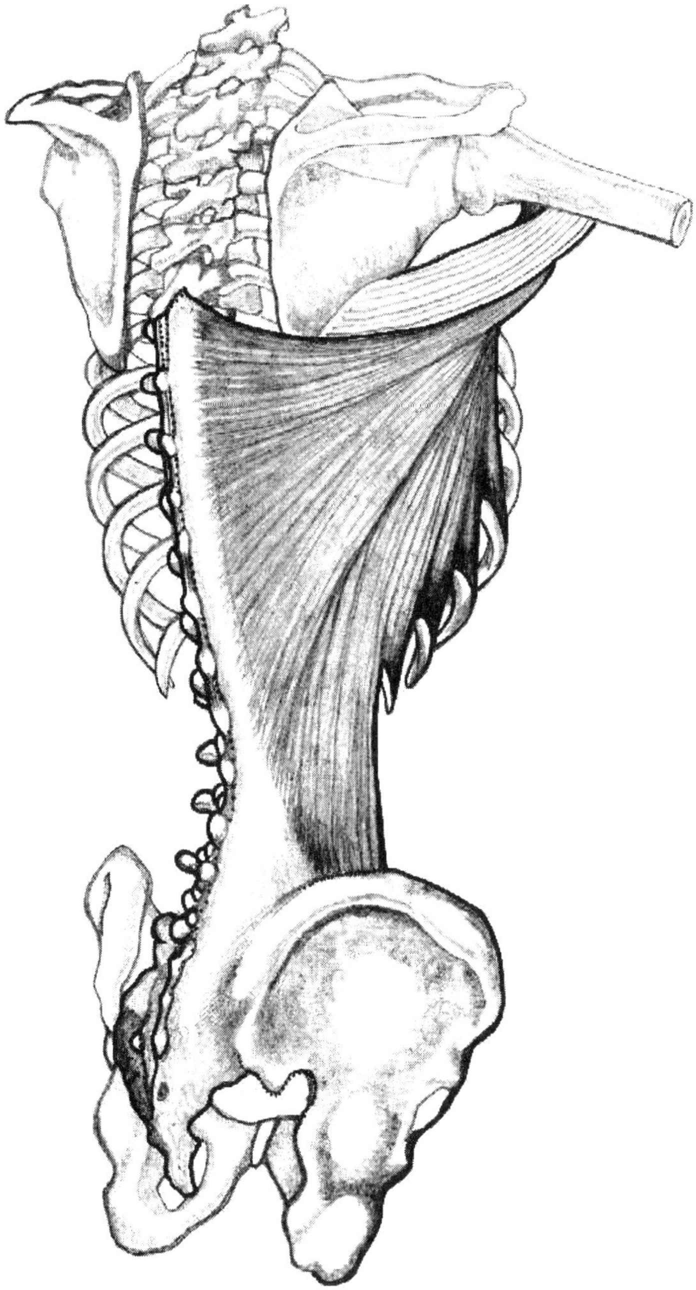

Origin:
T-6 through T-12 vertebrae
(spinous processes via the thoracolumbar aponeurosis),
L-1 through L-5 and sacral vertebrae
(spinous processes),
Supraspinal ligament,
Ilium (posterior 1/3 of iliac crest),
Ribs 9 through 12,
Scapula (inferior angle)

Insertion:
Humerus (medial lip of the intertubercular groove),
Deep fascia of the brachium

Action:
Extension, adduction, and internal rotation of the humerus

Innervation:
Spinal nerves C-6 through C-8 (ventral rami)
Thoracodorsal nerve

Relationship:
This broad muscle covers most of the lower back. The superior fibers are nearly horizontal, the lower fibers are nearly vertical. The muscle fibers twist 180 degrees upon themselves and around the teres major muscle as they reach their attachment. Trapezius is most efficient in activities such as swimming, climbing, swinging, and walking with crutches. Rowing and pullover exercises are used to strengthen this muscle. Symptoms include primarily midthoracic back pain, which is made worse by repetitive pulling downward type activities. This pain must be differentiated from that of suprascapular nerve entrapment, C-7 radiculopathy, ulnar neuropathy, and bicipital tendinitis.

RHOMBOID
Minor & Major (left)

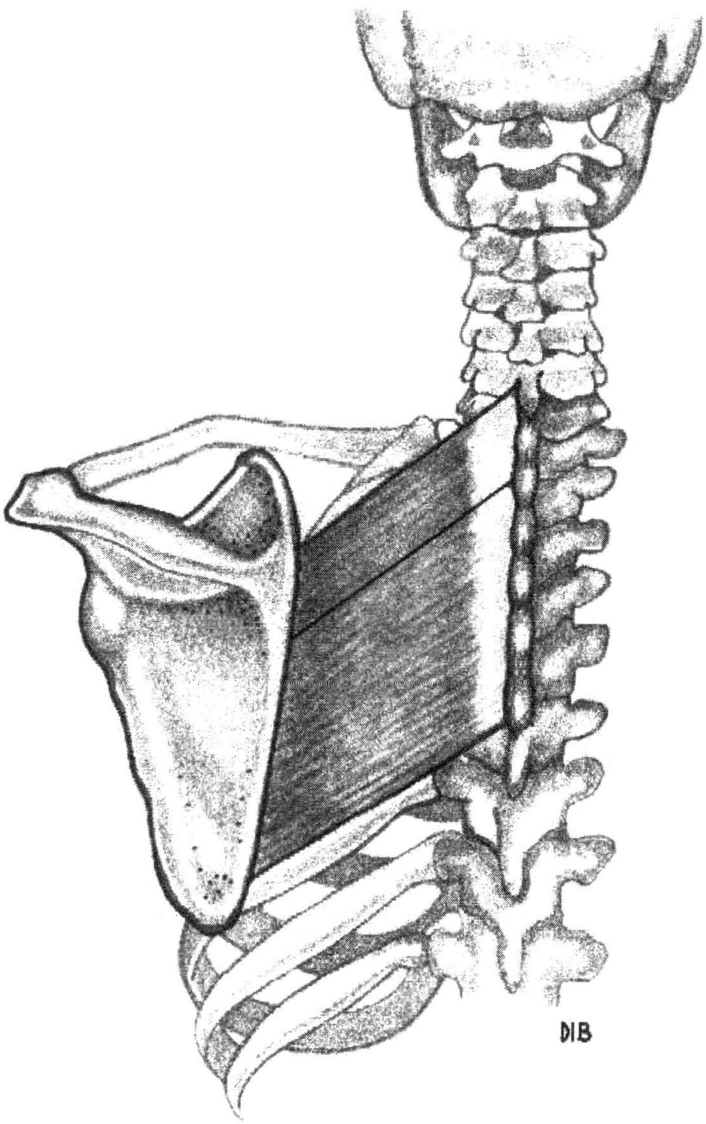

Origin:
Minor: Lower ligamentum nuchae and Vertebrae C-7 and T-1 (spinous processes)
Major: Vertebrae T-2 through T-5 (spinous processes & ligaments)

Insertion:
Minor: Medial border of the scapula at the root of spine
Major: Medial border of the scapula from the inferior root of spine to the inferior angle

Action:
Minor & Major: Scapular retraction (adduction) and downward rotation of glenoid fossa. Bilaterally, they extend the thoracic spine and assist in stabilizing the scapula

Innervation:
Minor & Major: Branch of dorsal scapular nerve, via upper trunk of the brachial plexus (portions of spinal nerves C-4 & C-5)

Relationship:
Both rhomboid major, a quadrilateral sheet of muscle and the smaller, cylindrical rhomboid minor, lie deep to trapezius and superficial to serratus posterior superior. They act synergistically with middle trapezius to oppose pectoralis major. Chin-ups and dips are useful exercises for strengthening these muscles. The rhomboids often exhibit lengthening, weakness, and pain when associated with hypertonic pectoral musculature (round-shouldered posture). Active trigger points of the rhomboids refer pain to the scapular region.

LEVATOR SCAPULA (left)

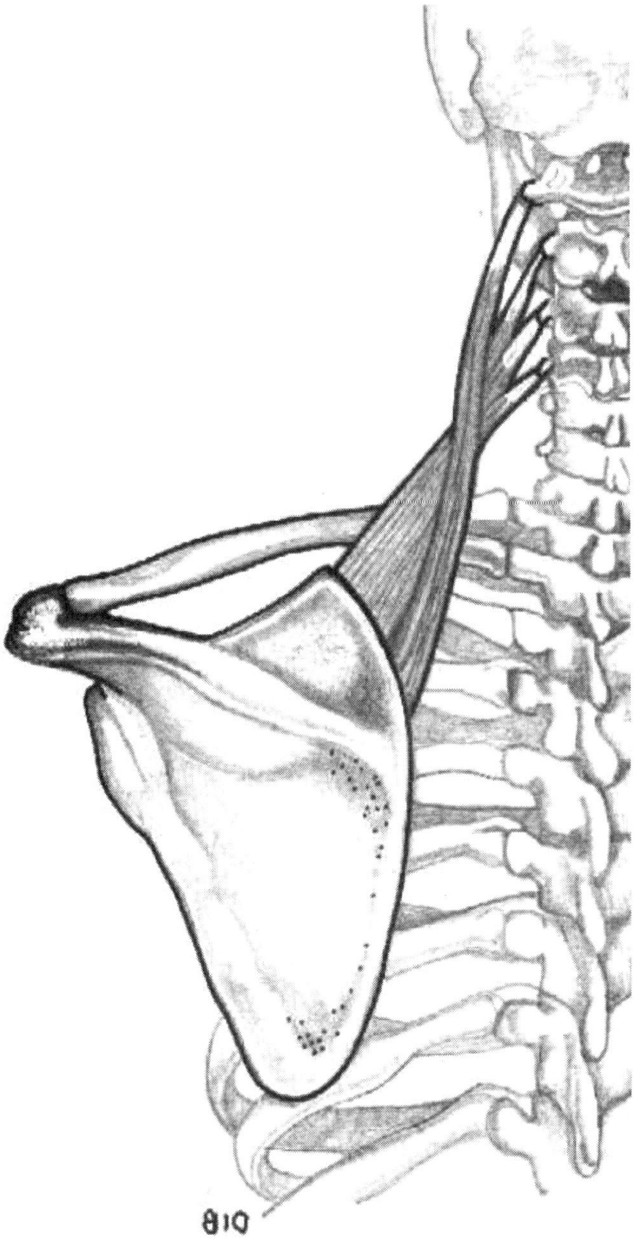

Origin:
Vertebrae C-1 & C2 (via tendinous slips to transverse processes) and C-3 & C-4 (via posterior tubercles of transverse processes). Accessory attachments are often found and are varied

Insertion:
Medial scapular border between the superior angle and root of spine of the scapula

Action:
Elevation of scapula (acting with upper trapezius), downward rotation of the glenoid fossa and (in concert with others muscles) stabilization of the scapula

Innervation:
Branches of spinal nerves C-3 & C-4 (and by C-5 via the dorsal scapular nerve)

Relationship:
This slender muscle twists as it moves from its origin to insertion. It lies deep to the trapezius muscle. This muscle is active in playing tennis, swimming, and spectator sports. Shoulder shrugs are useful exercises for strengthening this muscle. Stretching is most successful when accompanied by moist heat. Levator scapula is often involved in complaints of stiff neck and is associated with fatigue from occupational stress such as typing, telephone use, and carrying a handbag. Active trigger points of levator refer pain to the neck and vertebral border of the scapula.

SERRATUS ANTERIOR
(left) Lateral View

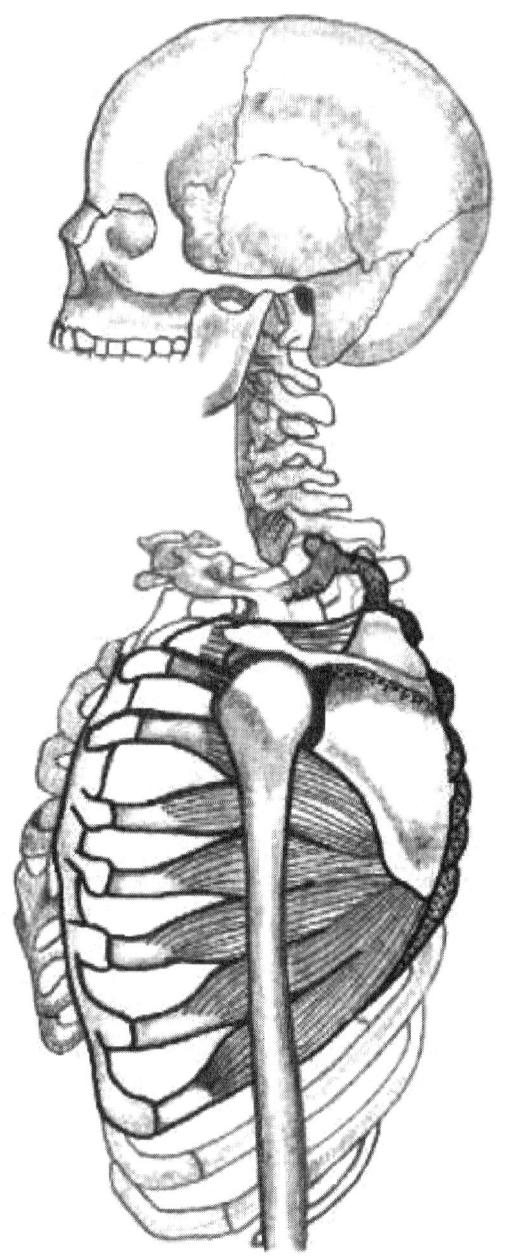

Origin:
Digitations arise anteriorly from the outer surfaces and superior borders of ribs 1 through 8 (occasionally also 9 & 10) and from intercostal fascia

Insertion:
Primary insertion is on the ventral (costal) surface of the vertebral border of the scapula from the superior angle to the inferior angle. Both the superior and inferior angles of the dorsal scapula may also provide sites for limited attachment

Action:
Protraction (abduction) of scapula, upward rotation of the glenoid fossa, and stabilization of the scapula against the posterior thorax during forward pushing efforts

Innervation:
Long thoracic nerve (from the anterior rami of C-5, C-6, C-7 & occasionally C-8 spinal nerves) *The long thoracic nerve lies superficial to the serratus anterior muscle

Relationship:
Serratus derives its name from the fingerlike (serrated) projections at its origin. It can be palpated along the anterior surface of the ribs during forceful scapular protraction. Serratus anterior is active in steering an automobile, swimming, playing tennis, shooting a basketball, throwing a baseball, and punching in boxing. Push-ups, bench press, and overhead press are useful exercises for strengthening this muscle. Paralysis or weakness of this muscle causes winging of the scapula. Active trigger points of serratus anterior may refer pain down the ulnar aspect of the arm and into the hand.

SPLENIUS CAPITUS (left)

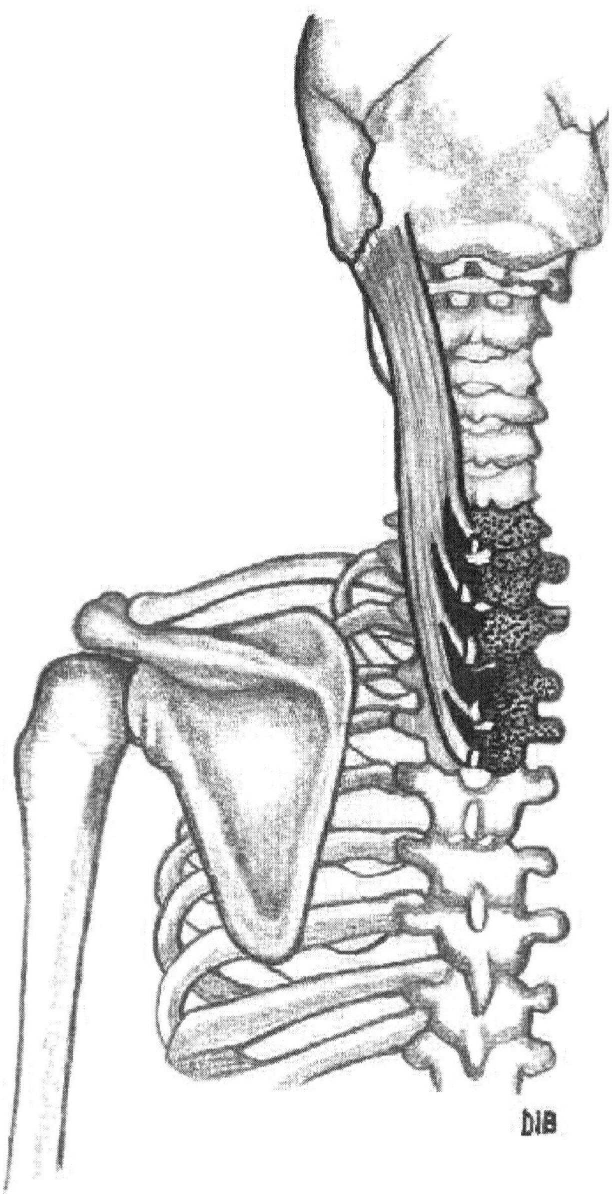

Origin:
Ligamentum nuchae (dorsal edge of lower half)
Vertebrae C-7 through T-3 (or T-4) (spinous processes and ligaments-deep to rhomboids & trapezius)

Insertion:
Mastoid process of the temporal bone and occipital bone (just below superior nuchal line & inferior to the attachment of SCM)

Action:
Unilaterally: rotation of the face/head to the same side
Bilaterally: extension of the head and neck (actions are synergistic to contralateral SCM)

Innervation:
Spinal nerves C-2 through C-4 (lateral branches of dorsal rami) and sometimes those of C-1 & C-5

Relationship:
The insertion of splenius capitus is deep to that of the sternocleidomastoid muscle. The term "splenius" comes from the Latin meaning bandage. Splenius appears to wrap the deeper muscles of the neck. Extension against resistance is a useful exercise for strengthening the splenius. This muscle is often involved in whiplash injuries and becomes vulnerable to dysfunction when overused and exposed to cold. Trigger point pain is referred primarily to the vertex of the head. Complaints must be differentiated from those of neurological disease, torticollis, and originating in other muscles.

SPLENIUS CERVICIS (left)

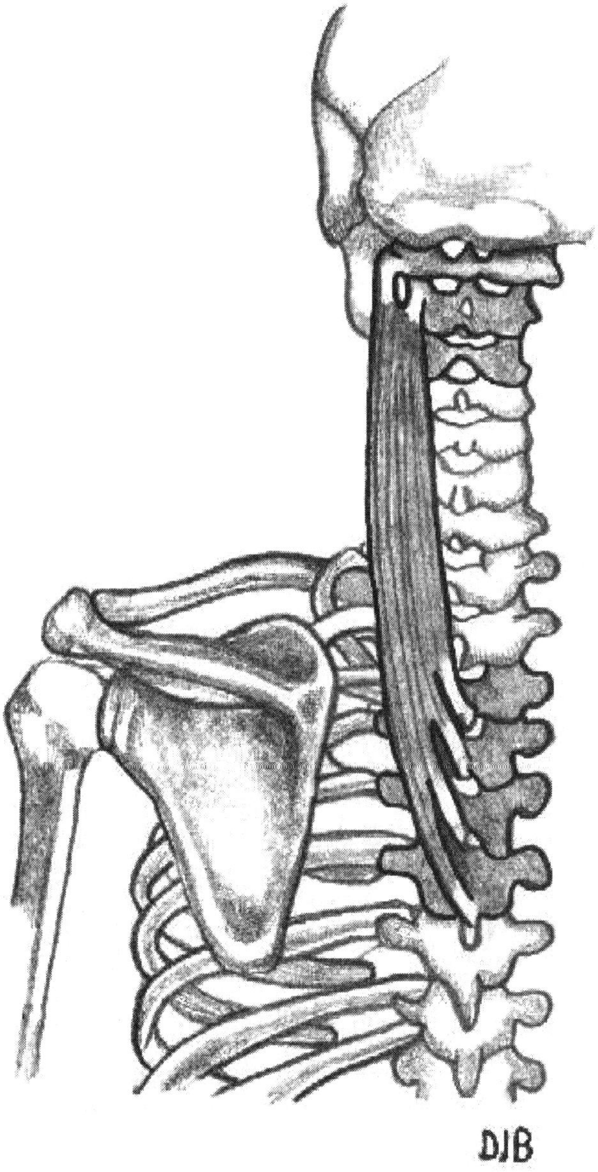

Origin:
Vertebrae T-3 through T-6 (spinous processes)

Insertion:
Vertebrae C-1 through C-2 (or C-3) (posterior tubercles of the transverse processes) On the posterior tubercles, splenius cervicis is slightly anterior to the attachment of levator scapula

Action:
Unilaterally: rotation of the face/head to the same side and lateral flexion (Sidebending) to the same side
Bilaterally: extension of the head and neck (its actions are synergistic to the contralateral SCM)

Innervation:
Spinal nerves C-3 through C-6 (lateral branches of the dorsal rami)

Relationship:
Splenius cervicis lies lateral and superior to splenius capitus. The paired cervicis form a "V" shape. Both splenius muscles lie deep to the trapezius and appear to wrap around other posterior cervical muscles. Extension against resistance is a useful exercise for strengthening the splenius. Like capitus, cervicis is often injured in whiplash and is commonly implicated with headaches. Cervicis is also susceptible to chronic tension, fatigue, and cold exposure. Trigger point pain may be referred to the occiput, the eye region, or inferiorly to the angle of the neck. Complaints must be differentiated from those of neurological disease, torticollis, and originating in other muscles.

SERRATUS POSTERIOR SUPERIOR (left)

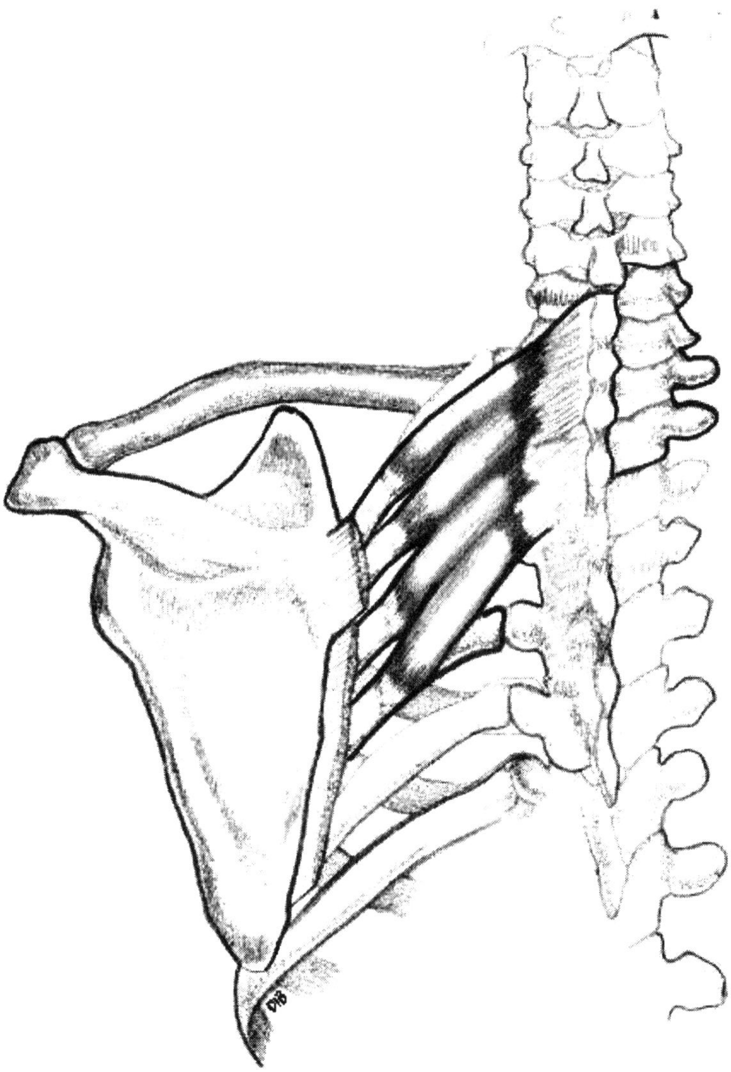

Origin:
Vertebrae C-7 through T-2 or T-3 (spinous processes & supraspinous ligaments), Ligamentum nuchae (via thin aponeurosis from lower portion)

Insertion:
Ribs 2 through 5 (via four digitations to the upper borders and external surfaces)

Action:
Elevates ribs, thereby increasing thoracic volume and aiding inspiration (although this action is debated by some authors, the muscle presumably acts synergistically with the diaphragm, intercostals, levator costarum, and scalenes in respiration)

Innervation:
Intercostal nerves 2 through 5 (which arise from anterior divisions of spinal nerves T-1 through T-4)

Relationship:
The serratus posterior superior muscle lies deep to the rhomboids and superficial to the thoracic portion of the thoracolumbar fascia. The digitations often vary in number from 3 to 6 and may even be absent in some specimens. This muscle may be clinically involved in scoliosis and may present with pain in patients with respiratory disease. Active trigger points of serratus posterior superior refer to an area deep to the superior portion of the scapula, posterior shoulder, upper triceps area, and ulnar aspect of forearm, hand, & little finger. Complaints must be differentiated from those of thoracic outlet syndrome, C-7/C-8 radiculopathy, olecranon bursitis, and ulnar neuropathy.

ERECTOR SPINAE (left)
Posterior View

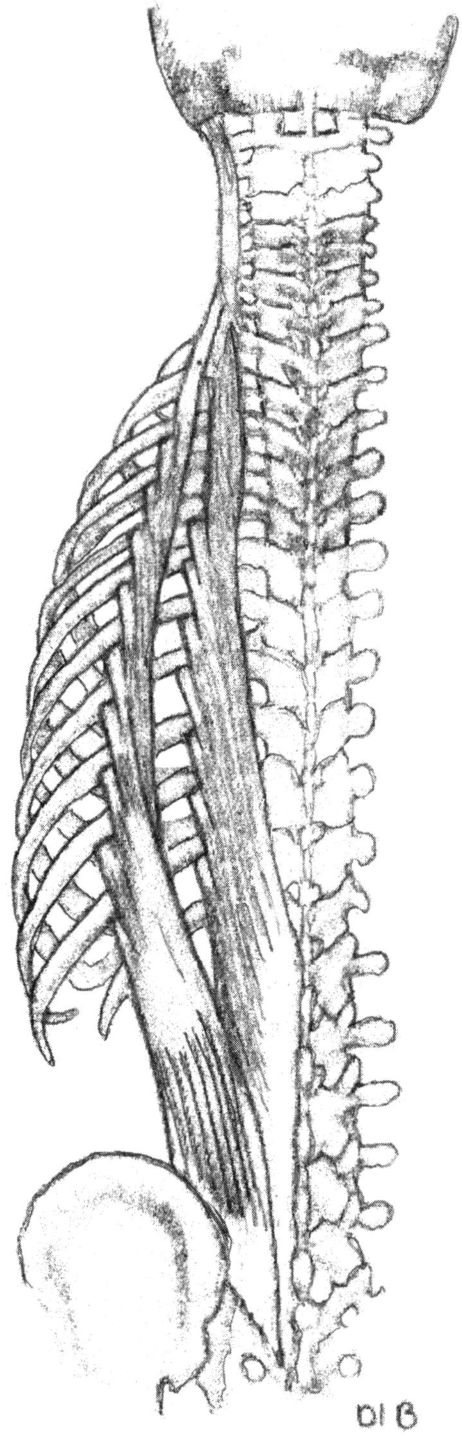

Origin:
Illiocostalis: Thoracolumbar fascia and posterior ribs
Longissimus: Thoracolumbar fascia, lumbar and thoracic transverse processes
Spinalis: Ligamentum nuchae, cervical and thoracic spinous processes

Insertion:
Illiocostalis: Posterior ribs and cervical transverse processes
Longissimus: Cervical and thoracic transverse processes, mastoid process
Spinalis: Cervical and thoracic spinous processes, occiput

Action:
Bilaterally: Extension of the trunk
Unilaterally: Lateral flexion of the trunk (sidebending) to ipsilateral side

Innervation:
Dorsal rami of spinal nerves

Relationship:
Located superficial to the transversospinalis, and moving from lateral to medial, iliocostalis is most lateral, longissimus is more medial, and spinalis is most medial. Where iliocostalis has attachments throughout the full length of the spinal column, longissimus and spinalis are found only in the cervical and thoracic regions. All can be palpated along the length of the spine. Although associated with posture, the erector spinae group can achieve complete relaxation while a subject is in a standing position. This muscle group tends to be associated with gross vertebral movement, while the transvospinalis group seems to be associated with fine vertebral adjustments. The erector spinae group can be strengthened by extension exercises against resistance (such as with properly performed dead lifts or on specific back extension equipment).

TRANSVERSOSPINALIS (right)
Posterior View

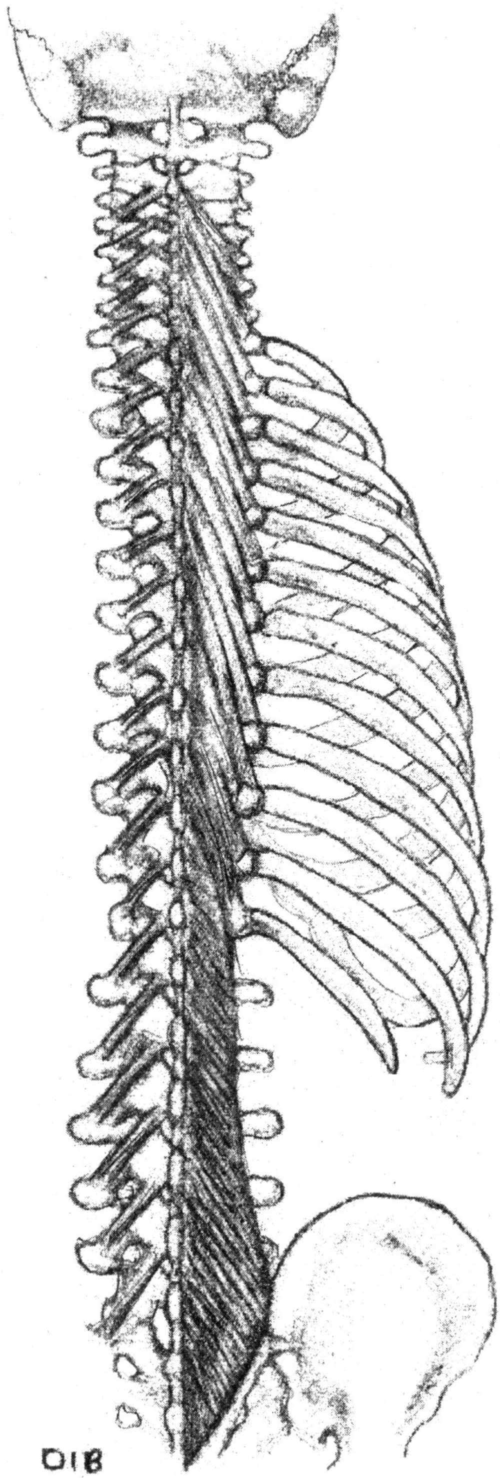

Origin:
Semispinalis: Cervical and thoracic transverse processes
Multifidus: Transverse processes of all vertebrae, sacrum & P.S.I.S.
Rotatores: Transverse processes of all vertebrae
Interspinales: Spinous processes of all cervical, T-1 & T-2 and T-10 to T-12 and all lumbar vertebra
Intertransversarii: Transverse processes of all cervical, T-10 to T-12 and all lumbar vertebrae

Insertion:
Semispinalis: Occiput, cervical and thoracic spinous processes of (3 to 6 superior) vertebrae
Multifidus: Spinous processes of all (2 to 4 superior) vertebrae
Rotatores: Spinous process of vertebra immediately below vertebra of origin
Interspinales: Spinous process of vertebra immediately below vertebra of origin
Intertransversearii: Transverse process of vertebra immediately below vertebra of origin

Action:
Semispinalis, Multifidus & Rotatores: Unilaterally they rotate the trunk to the contralateral side. Bilaterally they extend the trunk
Interspinales: Extension of trunk
Intertransversarii: Flexion of spine to ipsilateral side

Innervation:
Dorsal rami of spinal nerves

Relationship:
The transversospinalis group is deep to erector spinae and is in order from superficial to deep: semispinalis, multifidus, rotatores, interspinales, and intertransversarii. Superficial muscles have longer spans, while deeper muscles span only a few vertebrae.

Interspinales

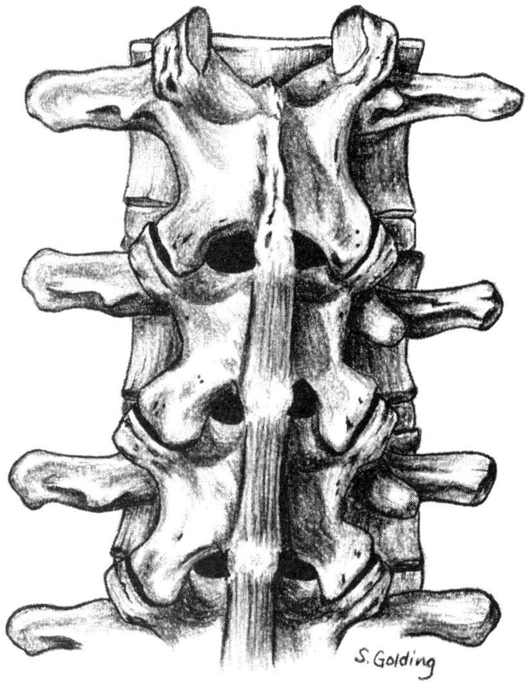

Origin:
Spinous processes of all cervical, T-1 and T-2, T-10 to T-12, and all lumbar vertebra

Insertion:
Spinous process of vertebra immediately below vertebra of origin

Action:
Extension of trunk

Innervation:
Dorsal rami of spinal nerves

Relationship:
Interspinales is deep to the rotatores and superficial to the intertransversarii.

Intertransversarii

Origin:
Transverse processes of all cervical, T-10 to T-12 & all lumbar vertebrae

Insertion:
Transverse process of vertebra immediately below vertebra of origin

Action:
Flexion of spine to ipsilateral side

Innervation:
Dorsal rami of spinal nerves

Relationship:
Intertransversarii is deep to the Interspinales.

TERES MAJOR (left)
Posterior View

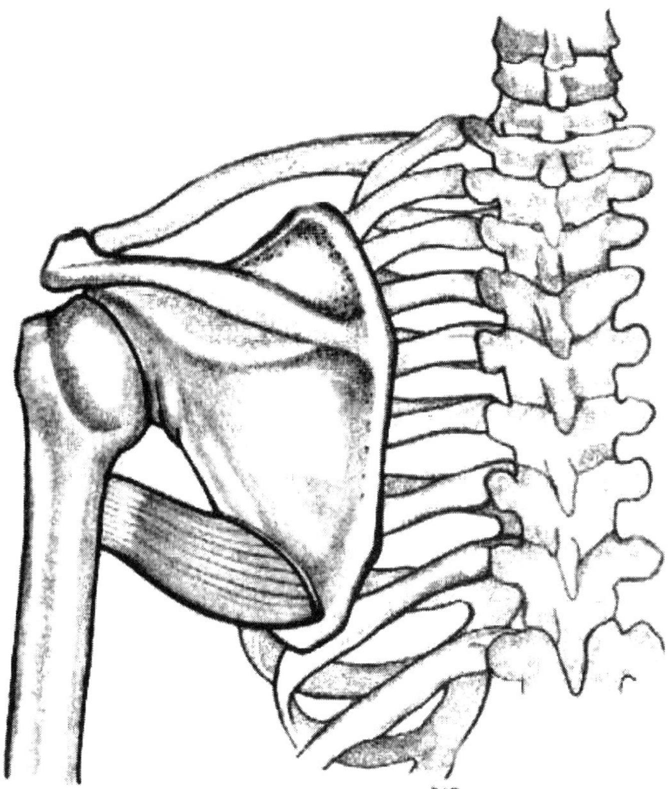

Origin:
Dorsal surface of the scapula near the inferior angle and from the fibrous septa shared with teres minor and infraspinatus muscles

Insertion:
Medial lip of bicipital groove of the humerus (via a flat tendon). The tendonous attachment site may be shared with latissimus dorsi or be slightly posterior and separated by a small bursa

Action:
Assists with internal rotation, (forceful) adduction and extension (from a flexed position) of the humerus

Innervation:
Lower subscapular nerve (via posterior cord from spinal nerve roots C-5, C-6, & C-7)

Relationship:
Teres major is a flat, thick muscle whose actions are similar to the latissimus dorsi. Both muscles contribute to the posterior axillary border. The tendons of teres major and latissimus pass between the triceps brachii and coracobrachialis muscles. Teres major may be strengthened with rope climbing, lat pulls, and forceful internal rotation against resistance. This muscle is stressed by repetitive overhead lifting, tennis (serving), and turning of an auto steering wheel. Pain on motion is the most common symptom and should be differentiated from disc disease, thoracic outlet syndrome, bursitis, and supraspinous tendonitis. Trigger point pain may be referred to the posterior deltoid region and down the dorsal aspect of the arm.

SERRATUS POSTERIOR INFERIOR (left)

Posterior View

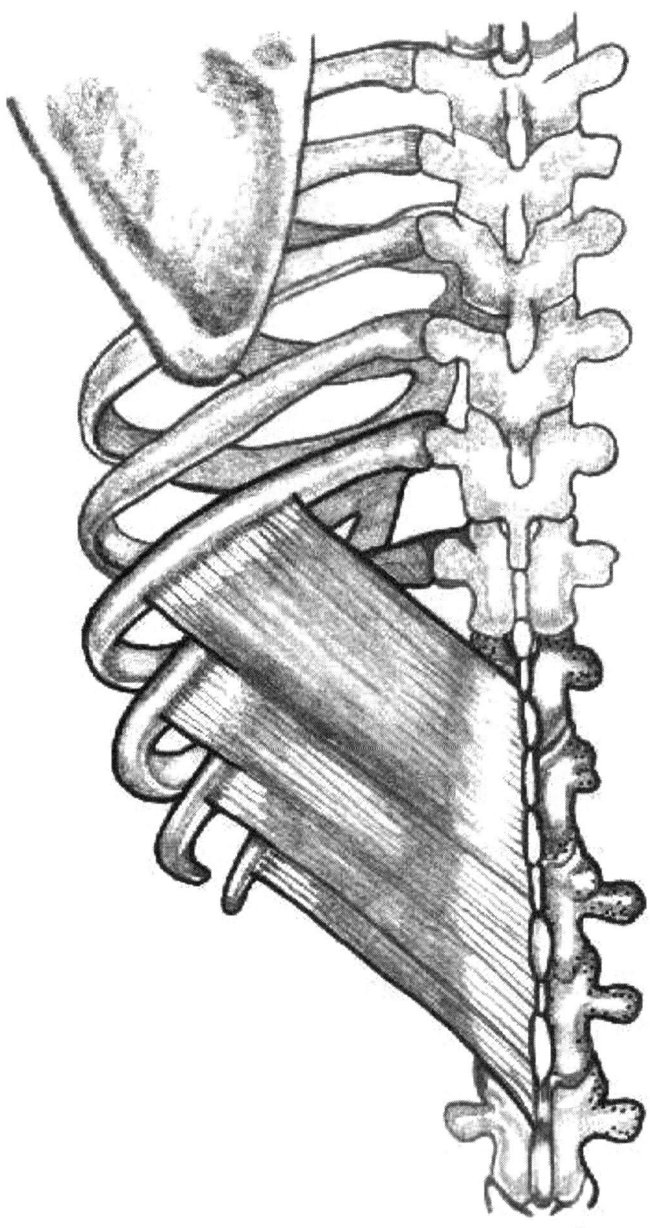

Origin:
Vertebrae T-11, T-12, L-1 and L-2 (spinous processes via aponeurosis)
Supraspinous ligaments

Insertion:
Ribs 9 through 12 (inferior & lateral borders, near to the angles of the ribs)

Action:
Unilaterally: trunk rotation to the ipsilateral side
Bilaterally: trunk extension and (possibly) stabilize lower ribs during exhalation (to counteract the superior pull of the diaphragm)

Innervation:
Intercostal nerves via spinal nerves T-9 through T-12 (anterior primary divisions only)

Relationship:
This thin, irregularly quadrilateral shaped muscle is located deep to latissimus dorsi. Fibers of the muscle run superiolaterally from the border between the thoracic and lumbar regions to just below the inferior angle of the scapula. Serratus posterior inferior is commonly aggravated by a combination of lifting, turning & reaching activities, and often presents with a chronic nagging ache. Symptoms of this muscle might mimic those of local disc related neuropathies or renal disease.

QUADRATUS LUMBORUM
(left) Anterior View

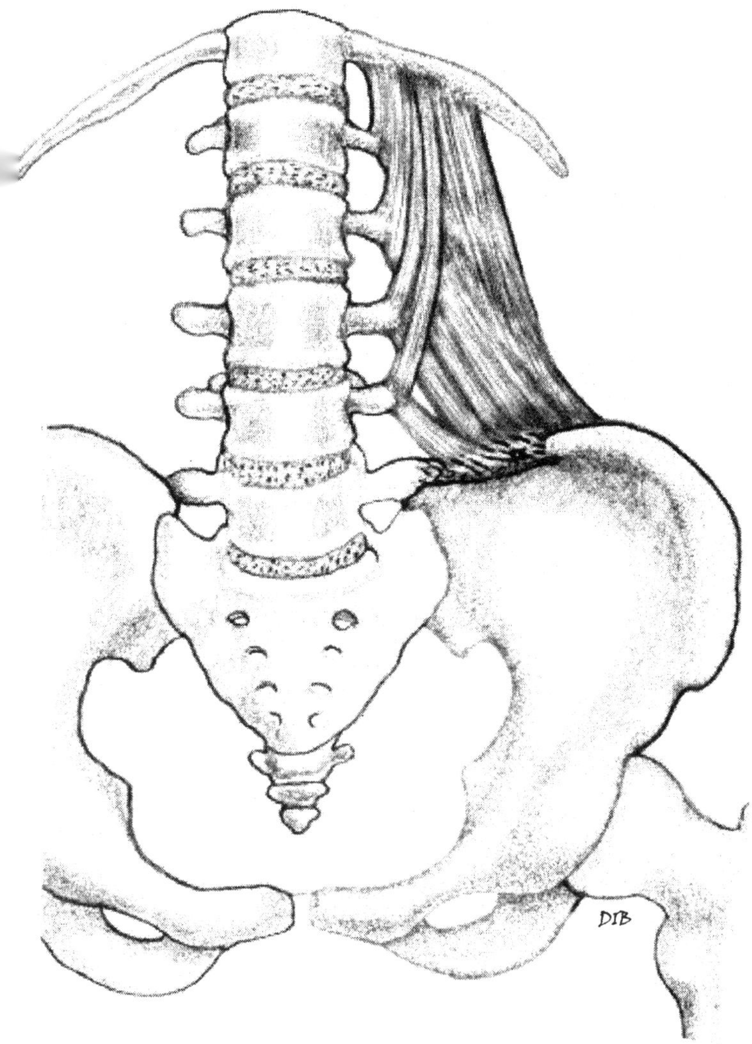

Origin:
Posterior iliac crest (approx. 5 cm)
Iliolumbar ligament (via aponeurosis)

Insertion:
Rib 12 (medial half of lower border)
Vertebrae L-1 through L-4 (transverse processes via small tendons)

Action:
Unilaterally: Lateral flexion of the spine (with pelvis fixed) and elevates ipsilateral hip (with spine fixed)
Bilaterally: Extension of lumbar spine
Stabilization of the 12th rib (during inspiration)

Innervation:
Spinal nerves T-12 & L-1 through L-4 (via branches of the lumbar plexus)

Relationship:
This irregularly quadrilateral shaped muscle which is wider inferiorly, is known as the "hip hiker" muscle. This muscle is involved in lifting, bending w/twisting, and rising from a seated position. Useful exercises to strengthen this muscle include trunk rotation and lateral flexion against resistance. Quadratus Lumborum is commonly associated with low back pain (deep & aching) that refers into the sacro-iliac joint and gluteal muscles. Trigger point pain may be referred to the hip and groin regions. Symptoms are common following auto accidents, repetitive strains, and walking/running on a slanted surface. Symptoms could be confused with those of SI joint dysfunction, trochanteric bursitis, disc disease, or a plethora of internal maladies.

SUPRASPINATUS (left)
Posterior View

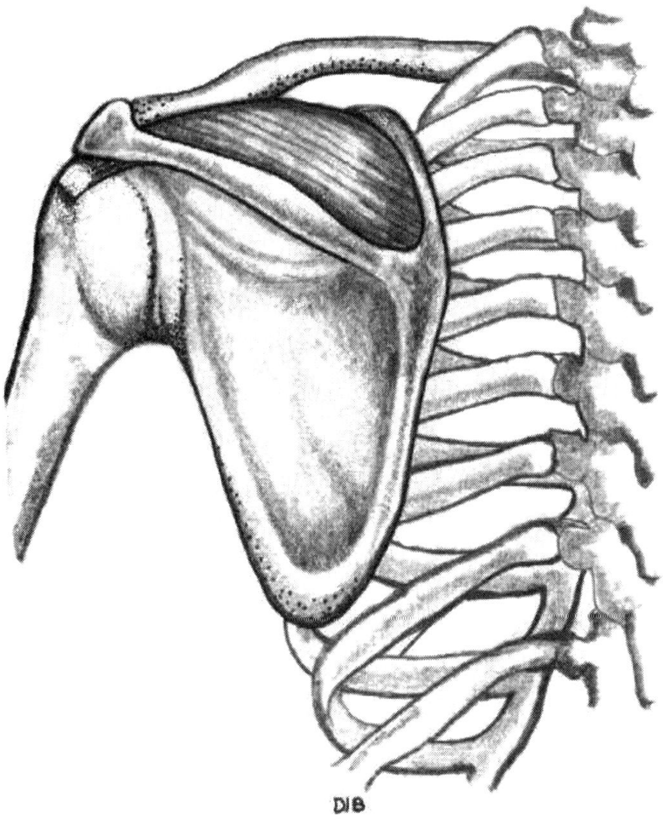

Origin:
Medial 2/3 of supraspinous fossa of the scapula and from the supraspinous fascia

Insertion:
Greater tubercle of humerus (superior facet) and to the superior and posterior portion of the joint capsule

Action:
Abduction of the humerus. Stabilizes head of the humerus within the glenoid fossa of scapula during forceful arm movement, thus preventing inferior displacement of the humerus

Innervation:
Suprascapular nerve (via upper trunk of brachial plexus from spinal nerves C-5 & C-6)

Relationship:
It was previously postulated that supraspinatus might be responsible for initiating humeral abduction, but now evidence supports this muscle and deltoid working as a team (see "scapulohumeral rhythm" under deltoid.) This muscle is used in activities such as carrying heavy objects, working overhead, and working with arms elevated. Supraspinatus is the most easily fatigued shoulder muscle and thus subject to overuse injury. The tendon of supraspinatus is separated from adjacent structures by the large subacromial bursa that can be a source of pathology. The tendon of supraspinatus is the most commonly injured component of the rotator cuff. Supraspinatus symptoms may mimic those of brachial plexus syndrome, cervical arthritis, or subdeltoid bursitis. Trigger point pain may be referred to deep in the shoulder and down the arm.

INFRASPINATUS (left)
Posterior View

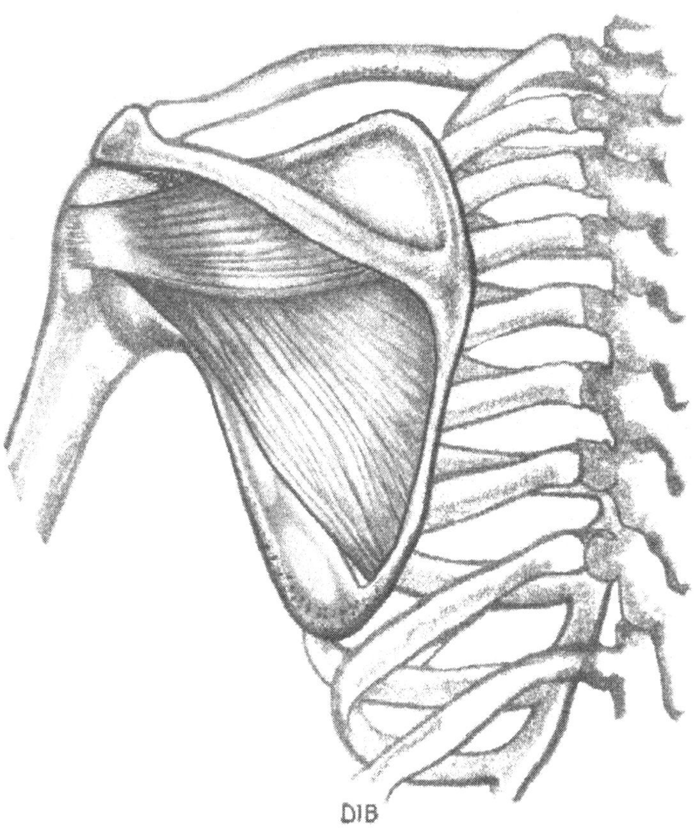

Origin:
Medial 2/3 of the infraspinous fossa of the scapula and the adjacent fascia, which separates it from teres minor

Insertion:
Posterior aspect of greater tubercle of humerus (middle facet) and to the superior & posterior portion of the joint capsule

Action:
External rotation of the humerus.
Stabilizes head of humerus within the glenoid fossa of the scapula during forceful arm movement (particularly preventing inferior displacement of the humerus)

Innervation:
Suprascapular nerve (via upper trunk of brachial plexus from spinal nerves C-5 & C-6)

Relationship:
Infraspinatus is a thick, triangular muscle occupying most of the infraspinous fossa of the scapula. It lies deep to the deltoids, trapezius, and latissimus dorsi. This muscle is most active in unusual movements such as reaching (out and posteriorly) and in the deceleration of movements initiated by more powerful anterior muscles. Chin-ups, rope climbing, and dips strengthen this muscle. Infraspinatus is commonly traumatized by acute overload rather than by prolonged overuse. Infraspinatus symptoms might include local pain, an inability to simultaneously adduct, extend, and medially rotate, or a weakness of handgrip. Symptoms may mimic those of disc disease, bicipital tendonitis, or arthritis. Trigger point pain may be referred to the shoulder and down the arm.

TERES MINOR (left)
Posterior View

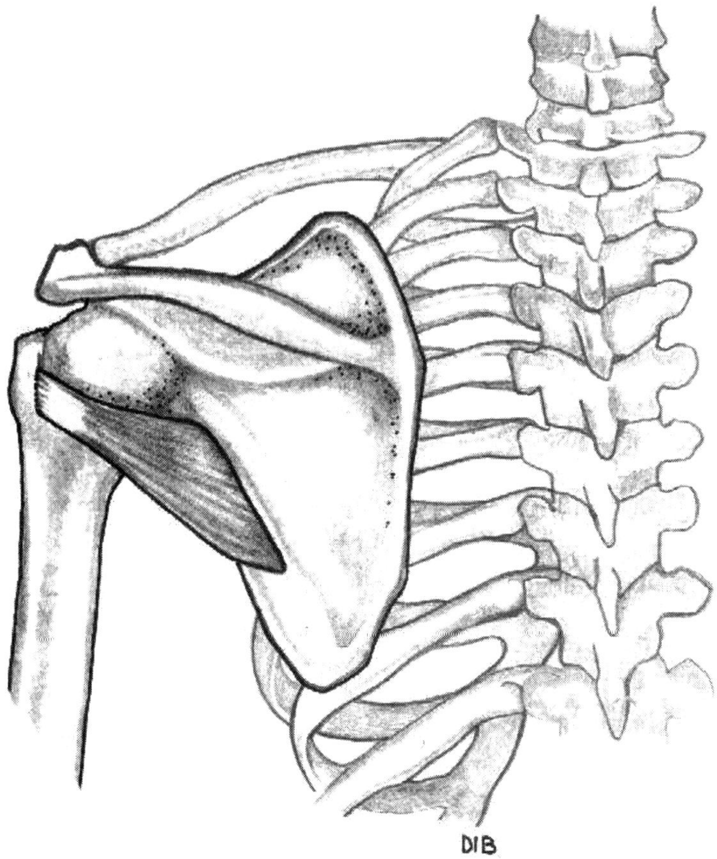

Origin:
Superior 2/3 of axillary border of the scapula (posterior surface) and to the aponeurosis separating it from infraspinatus and teres major

Insertion:
Greater tubercle of anterior humerus (inferior facet), to the posterior portion of the glenohumeral joint capsule and to the neck of the humerus, directly below the inferior facet

Action:
External rotation of the humerus. Stabilizes the head of the humerus within the glenoid fossa of the scapula, during forceful arm movement

Innervation:
Axillary nerve (via posterior cord of brachial plexus from spinal nerves C-5 & C-6)

Relationship:
Teres minor is a narrow elongated muscle. Its fibers run parallel to infraspinatus and obliquely upwards across the posterior shoulder joint. This muscle lies superior to teres major and deep to infraspinatus. Teres minor is involved in reaching or lifting overhead, playing volleyball, and driving an automobile. It may be strengthened by chin-ups, rope climbing, and dips. Teres minor symptoms often include posterior shoulder pain with movement but usually not restricted motion and may mimic ulnar neuropathy, bursitis, or acromioclavicular separation. Trigger point pain may be referred to the deltoid region.

SUBSCAPULARIS (left)
Anterior View

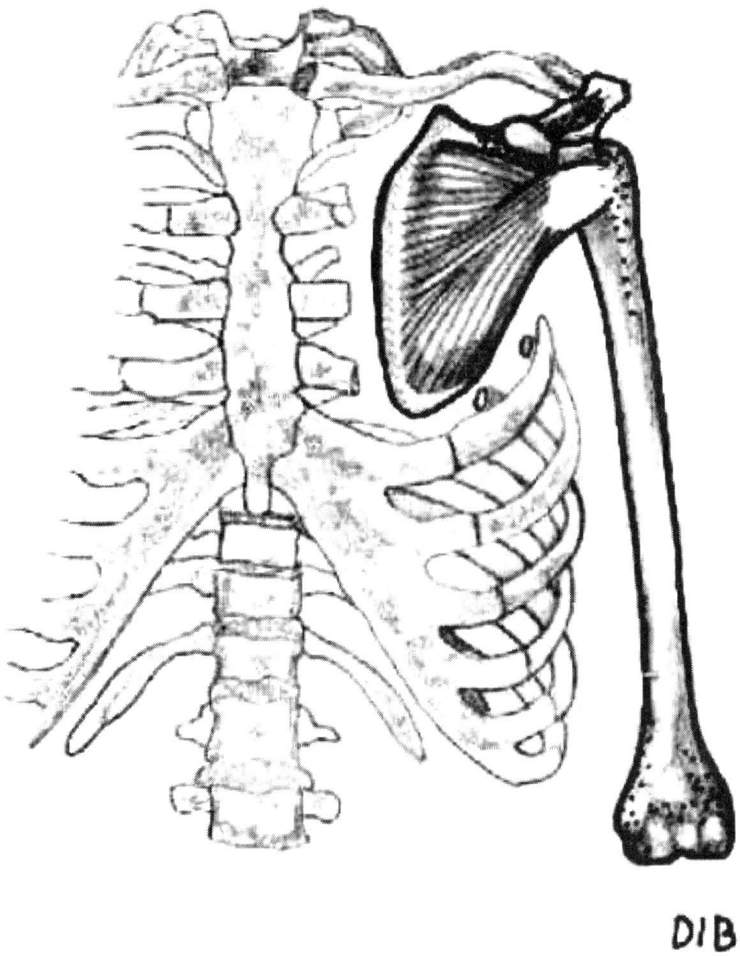

Origin:
Subscapular fossa (anterior surface) of the scapula, the groove along inferior 2/3 of the axillary border and the aponeurosis separating it from teres major and long head of the triceps brachii

Insertion:
Lesser tubercle of anterior humerus, lower portion of glenohumeral joint capsule & the neck of the humerus below the tubercle

Action:
Internal rotation and adduction of the humerus. Stabilization of the head of the humerus within the glenoid fossa & counteracting upward forces

Innervation:
Superior and inferior subscapular nerves (via posterior cord of brachial plexus from spinal nerves C-5 and C-6)

Relationship:
Subscapularis is a large triangular muscle covering the anterior surface of the scapula and forming much of the posterior axillary wall. It lies superficial to portions of the serratus anterior muscle and the large subscapular bursa. This muscle plays a prominent role in scapular mobility and is often involved with frozen shoulder. It may be strengthened by rope climbing, rowing, and lat pulls. Symptoms associated with subscapularis may mimic adhesive capsulitis, thoracic outlet syndrome, or impingement syndromes. Trigger point pain may be referred to the posterior deltoid region, triceps region, and down the arm.

DELTOID (left)
Lateral View

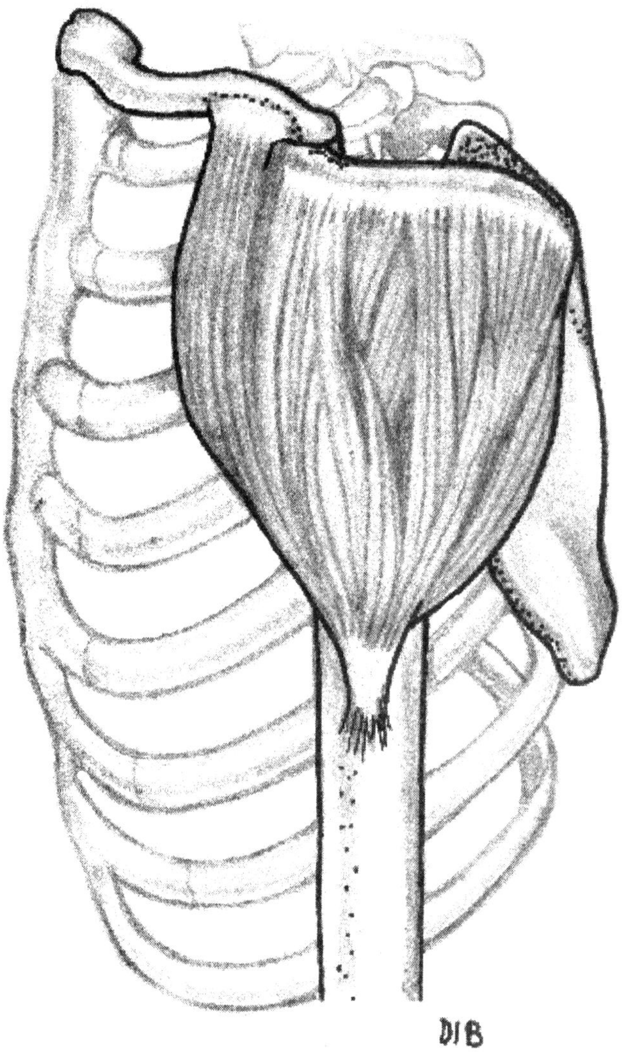

Origin:
Anterior: Lateral 1/3 of the clavicle (anterior border & superior surface)
Middle: Acromion process (lateral margin & superior surface)
Posterior: Spine of the scapula (lower edge of the crest)

Insertion:
Deltoid tuberosity of humerus

Action:
Anterior: Flexion, horizontal adduction and possibly assists with internal rotation of the humerus
Middle: Abduction of humerus
Posterior: Extension of humerus and possibly assists with external rotation of the humerus

Innervation:
Axillary Nerve (via posterior roots of spinal nerves C-5 & C-6)

Relationship:
The Deltoid is a thick, curved, triangular muscle. All three portions of it assist supraspinatus with abduction of the humerus in a complex process known as "scapulohumeral rhythm" (with maximal contractile forces of both muscles being generated when the humerus is being raised from 90 to 180 degrees), and with stabilization of the head of the humerus within the glenoid fossa. Dumbbell raises or "flys" are useful exercises for strengthening the deltoid muscle. Trauma or overuse of the deltoid often presents with symptoms similar to rotator cuff injuries, bursitis, or acromioclavicular dysfunction. Trigger point pain may be referred down the lateral aspect of the arm.

TRICEPS BRACHII & ANCONEUS (left)
Posterior View

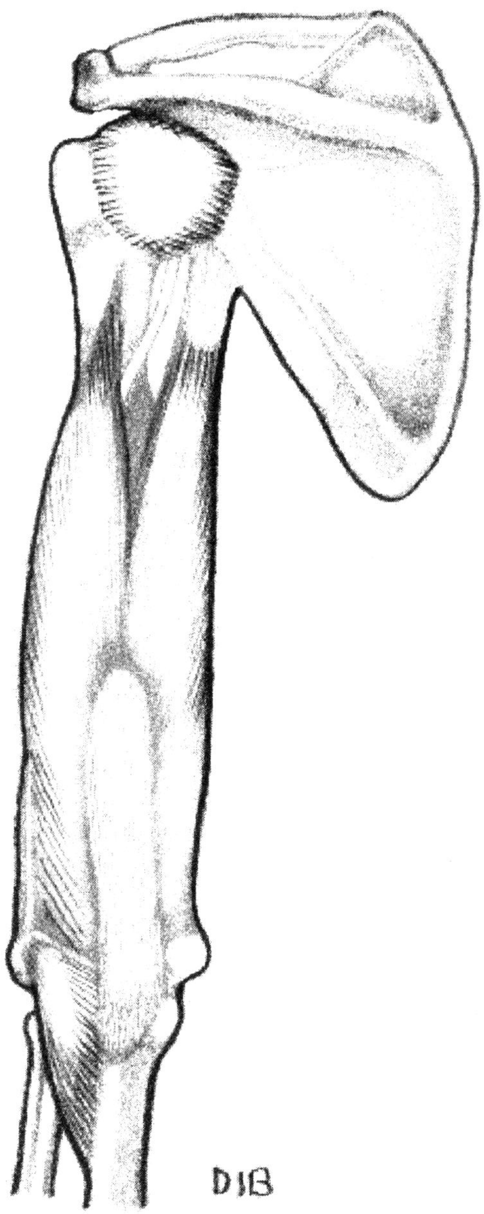

Origin:
Long head: Infraglenoid tubercle and lip of scapula
Lateral head: posterior surface of humerus proximal to the radial groove (and nerve)
Medial (deep) head: posterior surface of humerus distal to the radial groove (and nerve)
Anconeus: Lateral epicondyle of humerus

Insertion:
Olecranon process of ulna (via common tendon) & beyond to the antebrachial fascia
Anconeus: Lateral Olecranon process and proximal dorsal surface of ulna

Action:
Extension of the forearm at the elbow (the lateral & long heads are primarily active in extension against resistance). The long head may assist in extension of the humerus from the flexed position and stabilizing the head of the humerus within the glenoid fossa
Anconeus: Assists in extension of the forearm at the elbow and stabilizing the ulna during pronation or supination of the forearm

Innervation:
Radial nerve (via posterior cord from spinal nerves C-6, C-7 & C-8)

Relationship:
The name "triceps" means 3 heads. The Anconeus is often described as a distal & lateral extension of the triceps. Both muscles are used in activities such as walking with crutches, driving, throwing, jumping, playing tennis, golf, and baseball. Overhead presses, dips, pushups, and triceps extensions are useful for strengthening this muscle. Overuse injuries of the triceps are often associated with doing push-ups, driving a manual shift auto, and repetitive occupational activities. Triceps-related disorders might be confused with arthritis, bursitis, or thoracic outlet syndrome. Trigger point pain may be referred down the dorsal aspect of the arm.

GLUTEUS MAXIMUS
Posterolateral View

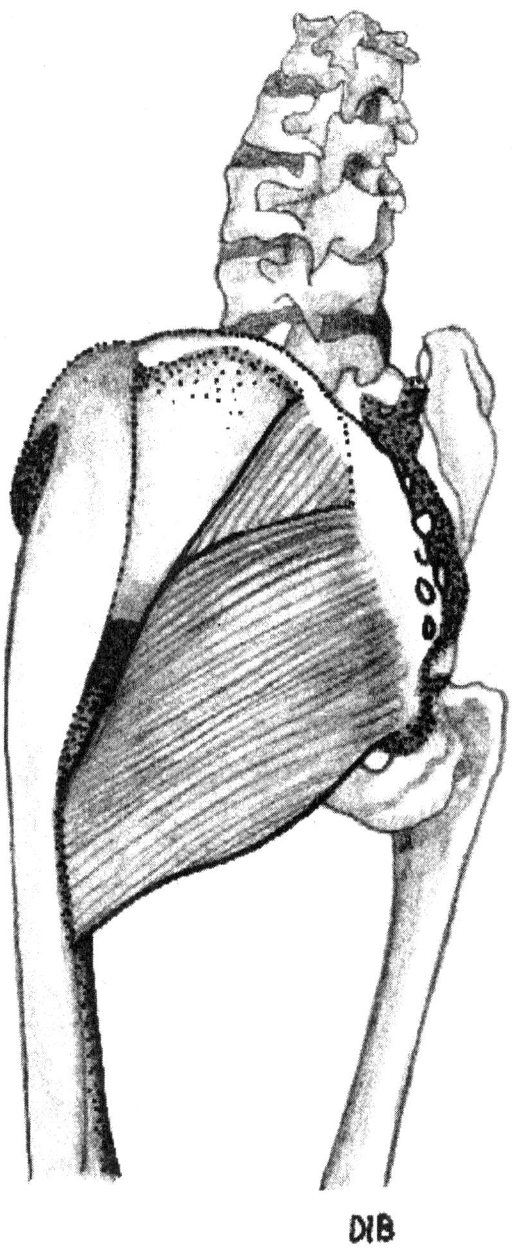

Origin:
Posterior iliac crest and posterior gluteal line, aponeurosis of erector spinae, lateral sacrum (dorsal surface), coccyx, sacrotuberous ligament, and fascia of gluteus medius

Insertion:
Greater trochanter of femur, iliotibial tract of tensor fascia latae, and the gluteal tuberosity of the femur

Action:
Extension of the flexed femur -- particularly past 15° (during strenuous activities such as: jumping, running, climbing stairs, & arising from a seated position). Assists in later rotation and abduction of femur. Stabilizes the pelvis during bending, standing, and walking

Innervation:
Inferior Gluteal Nerve (L-5, S-1, & S-2 spinal nerve roots)

Relationship:
Gluteus maximus is a broad, thick, anti-gravity muscle, with coarse fibers that are well suited for extended use. This muscle is uniquely developed in humans to allow for bi-pedal activities and upright posture. It forms the most superficial and prominent portion of the buttock. A large bursa separates this muscle from deeper structures. Weighted squats and trunk extensions are useful in strengthening this muscle. Hypertonicity of the gluteus maximus will restrict hip flexion. Pain associated with gluteus maximus is often aggravated by walking uphill, prolonged sitting, forward bending, and swimming (crawl stroke). Trigger point pain may be referred to the entire gluteal region.

GLUTEUS MEDIUS
Posterolateral View

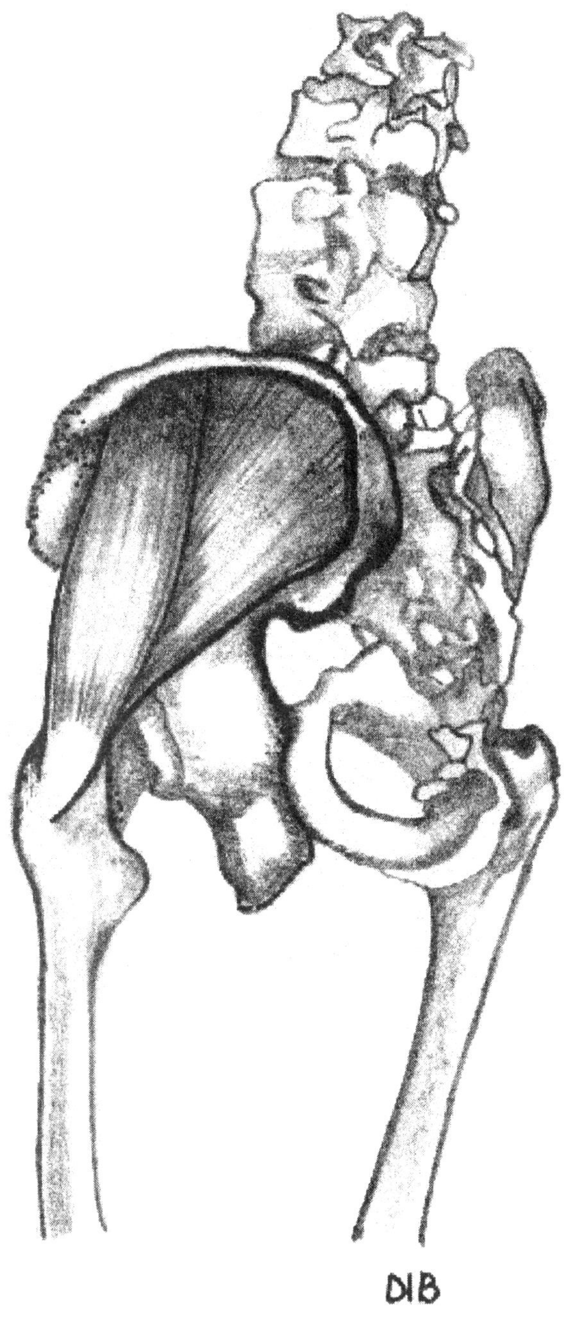

Origin:
Anterior ¾ of iliac crest, outer surface of ilium (between the anterior and posterior gluteal lines), and from strong superficial fascia

Insertion:
Posterior/superior angle and lateral surface of the greater trochanter of femur

Action:
Abduction of the femur, internal rotation of femur (anterior fibers)

Innervation:
Superior gluteal nerve, (spinal nerves L-4, L-5 & S-1)

Relationship:
Gluteus medius is a broad, thick, fan shaped muscle, which lies deep to the gluteus maximus (its posterior 1/3) and superficial to gluteus minimus. A bursa is found between its tendon and the greater trochanter. The primary function of gluteus medius is to stabilize the pelvis during ambulation, when body weight is shifted to the ipsilateral leg. Strengthening is best accomplished with side-lying leg raises or resisted leg abduction. Pain associated with gluteus medius is aggravated by running, extended walking on soft-uneven surfaces (sandy beach), tennis, and sudden falls. Pain may mimic that of sacroiliac origin. Hypertonicity of the gluteus medius will restrict adduction of the femur. Weakness or paralysis of this muscle will produce a pelvic tilt or drop on the contralateral side (Trendelenburg Sign). Gluteus medius trigger point pain may be referred to the lower back and the posteriolateral gluteal regions.

GLUTEUS MINIMUS
Posterolateral View

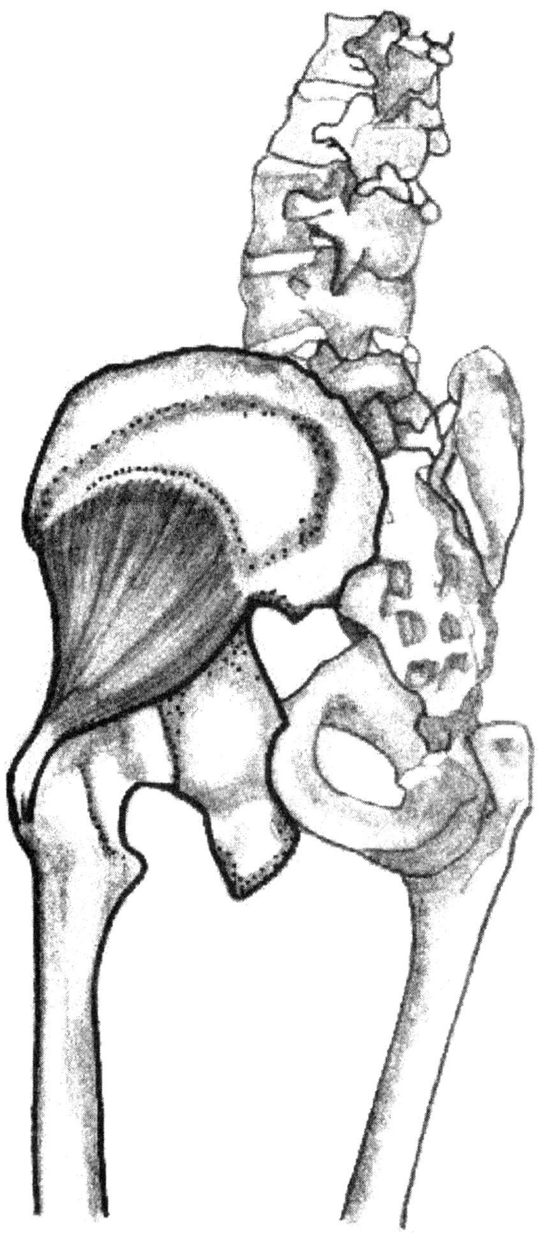

Origin:
Outer surface of ilium (between the anterior & inferior gluteal lines) and posteriorly near the margin of the sciatic foramen

Insertion:
Superior and anterior surface of the greater trochanter of femur (deep & anterior to the insertion of piriformis)

Action:
Abduction of the femur, internal rotation of femur (anterior fibers). Generally believed to assist gluteus medius in stabilizing the pelvis during ambulation

Innervation:
Superior gluteal nerve, (spinal nerves L-4, L-5 & S-1)

Relationship:
Gluteus minimus is the smallest and deepest of the gluteal muscles. Its fan shape conforms to that of the overlying gluteus medius, while its major function also corresponds to that related muscle. Strengthening is best accomplished by leg abduction exercises with resistance. Gluteus minimus problems may be aggravated by overuse (in running & tennis) or by walking (too fast or too far) on uneven ground. Gluteus minimus related pain might mimic that of gluteus medius, piriformis, (L-4, L-5 & S-1) disc related disease, or trochanteric bursitis. Those with gluteus minimus pain often exhibit an antalgic gait and are unable to cross the affected leg without pain. A weak gluteus minimus may contribute to finding a positive Trendelenburg Sign. Gluteus minimus trigger point pain may be referred to the inferiolateral gluteal region and down the lateral aspect of the thigh to the ankle.

TENSOR FACIAE LATAE
(left) Lateral View

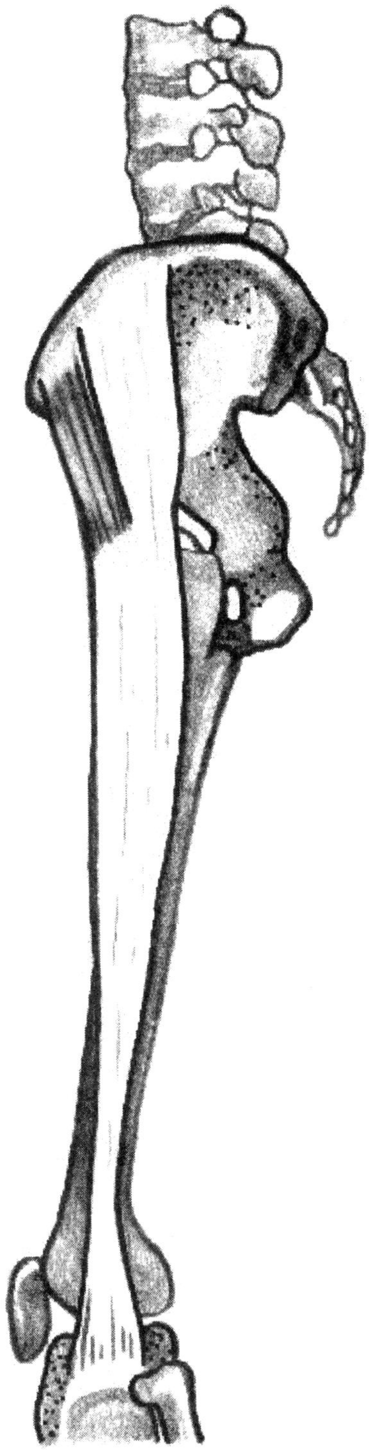

Origin:
Form the anterior portion of the outer lip of the iliac crest (near the iliac tubercle) posteriorly to the lateral portion of the anterior superior iliac spine. There may exist deep attachments to the superficial fasciae of gluteus medius

Insertion:
Anterolateral tibial condyle via the iliotibial tract

Action:
Acting through the iliotibial tract; abduction and flexion of the femur at the hip, and a tendency to rotate the femur internally as flexion occurs. Some authors debate the muscle's ability to abduct the femur, but add that it may extend the knee and laterally rotate the leg. It is generally believed that this muscle functions during ambulation to control rather than to produce movement

Innervation:
Superior gluteal nerve (via spinal nerves L-4, L-5, & S-1)

Relationship:
The tensor fasciae latae muscle descends between, and is attached to, two layers of the iliotibial tract. The muscle itself ends about 1/3 of the way down the femur, although it occasionally may extend as far as the lateral femoral condyle. This muscle helps to maintain the standing posture and (via the iliotibial tract) to stabilize the pelvis from below. It is a relatively small postural muscle and its actions are weak, but necessary, in directing the placement of the foot (in a sagittal plane) while walking, running, jumping, and cycling. Performing hip abductions against resistance strengthens tensor fasciae latae. Pain from the tensor fasciae latae is most often experienced as referred pain or tenderness in the area of the greater trochanter. Occasionally, pain is experienced down the lateral thigh to the knee. Pain is made worse by repetitive activities (walking, running, and cycling) and by prolonged sitting. Symptoms of tensor fasciae latae must be differentiated from those of trochanteric bursitis, lumbar neuropathy, arthritis, trigger points in other muscle, and iliotibial tract friction syndrome.

SIX DEEP LATERAL ROTATORS of the HIP

(Shown with Gluteals & Sciatic Nerve)
Posterior View

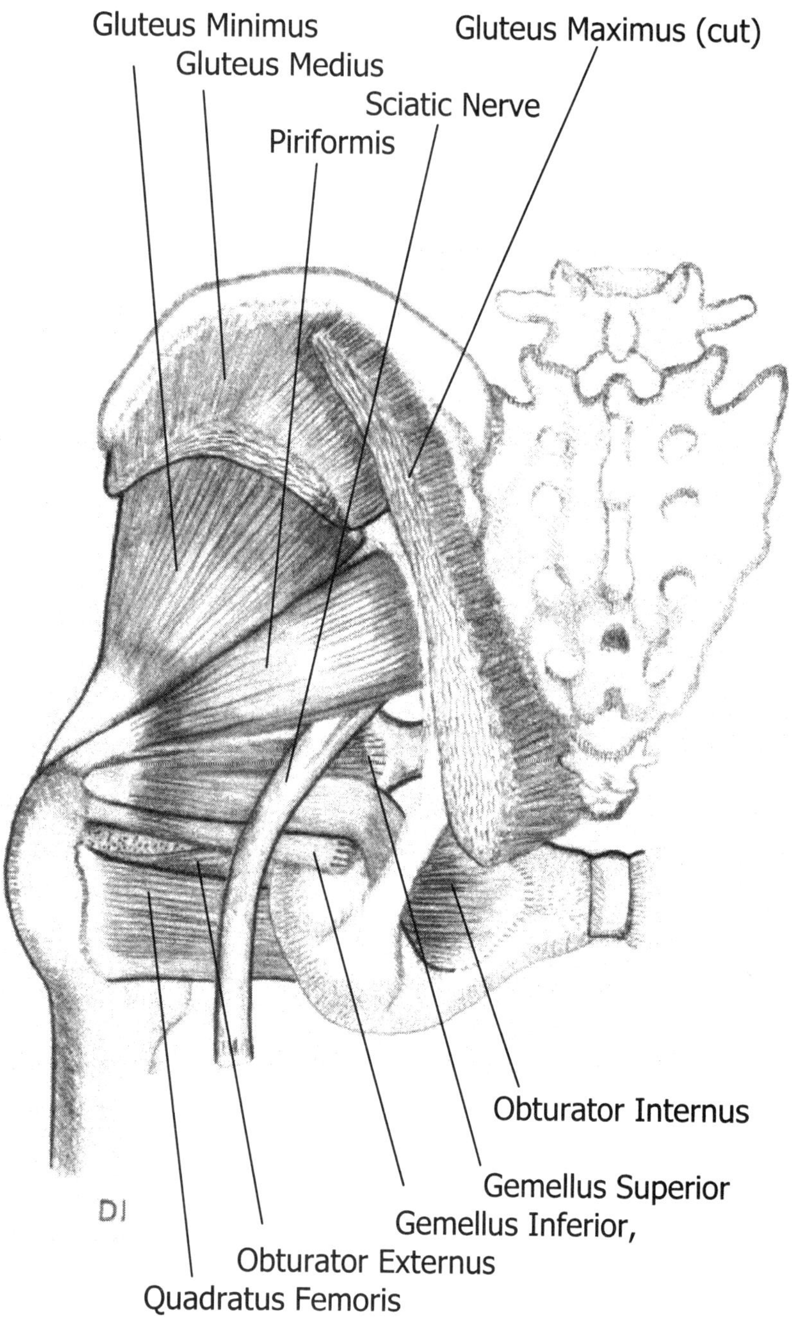

The six deep lateral rotators of the femur (hip) include: **P**iriformis, **G**emellus Superior, **O**bturator Internus, **G**emellus Inferior, **O**bturator Externus, and **Q**uadratus Femoris. They are shown here with gluteus minimus and dissected gluteus maximus & medius. Also shown is the sciatic nerve, which arises from the lumbosacral plexus (derived from spinal nerves T-12 thru S-3), exits from the pelvis thru the greater sciatic notch, and passes immediately deep to piriformis to enter the posterior thigh. These muscles all attach medially to the os coxae (inferior portions of the sacrum, ischium, ileum, & pubis), and travel laterally to terminate on the proximal femur (greater trochanter). As their name implies, they are located deep to the gluteal muscles and act synergistically with gluteus maximus to laterally (externally) rotate the femur. The pneumonic "**P**iece **G**oods **O**ften **G**o **O**n **Q**uilts" is often used to remember the order of these muscles (from superior to inferior).

PIRIFORMIS
Anterior View

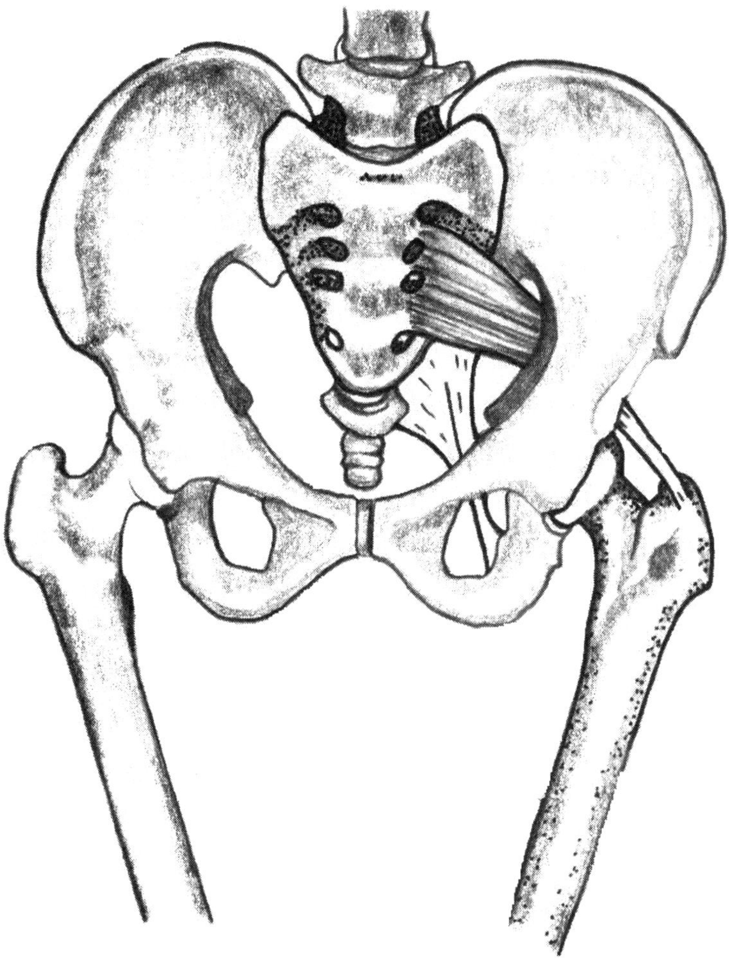

DIB

Origin:
Anterior surface of sacrum, anchored by three digitations to an area between the 1st & 4th sacral foramen (and occasionally to the medial margin of the sciatic foramen & the sacro-tuberous ligament)

Insertion:
Medial surface of the superior portion of the greater trochanter of femur (via a round tendon often shared with obturator internus & the gemelli muscles)

Action:
External rotation of femur (in non-weight bearing limb), Abduction of the femur (with hip flexed at 90 degrees)

Innervation:
Sacral Nerves (S-1 & S-2)

Relationship:
The piriformis is usually a thick and bulky, pear-shaped muscle (occasionally thin & rarely absent). It lies deep to the gluteals and medial to the other deep lateral rotators. It lies just posterior to the sciatic nerve, making piriformis capable of compressing the sciatic nerve (and vascular structures) against the rim of the greater sciatic foramen, causing "Piriformis Syndrome" with symptoms of sciatica. These symptoms mimic those of a herniated disc, facet syndrome, or SI joint dysfunction. Piriformis may be strengthened by external rotation against resistance. Piriformis trigger point pain may be referred to the SI region, entire gluteal region, and down the lateral aspect of the thigh.

GEMELLUS SUPERIOR
Posterior View

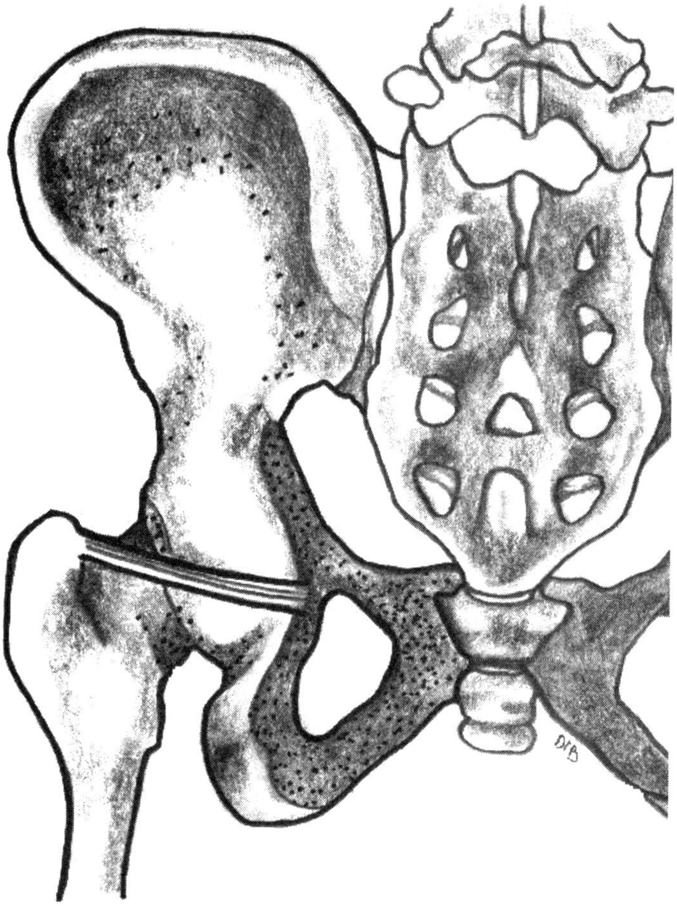

Origin:
Upper portion of ischial tuberosity

Insertion:
Medial surface of the greater trochanter (along with the tendon of obturator internus)

Action:
External rotation of femur (in the non-weight bearing, extended limb), Some authors also state that the gemelli may assist with abduction of the femur (with hip flexed at 90 degrees)

Innervation:
Obturator internus nerve (lumbar & sacral nerves L-5 & S-1)

Relationship:
Both gemelli may be considered as accessory muscles to obturator internus. This muscle may be absent in some specimens. The gemelli and obturator muscles are collectively referred to as the **GOGO** muscles. Tenderness deep to the gluteals and inferior to the piriformis may be attributed to the gemelli.

GEMELLUS INFERIOR
Posterior View

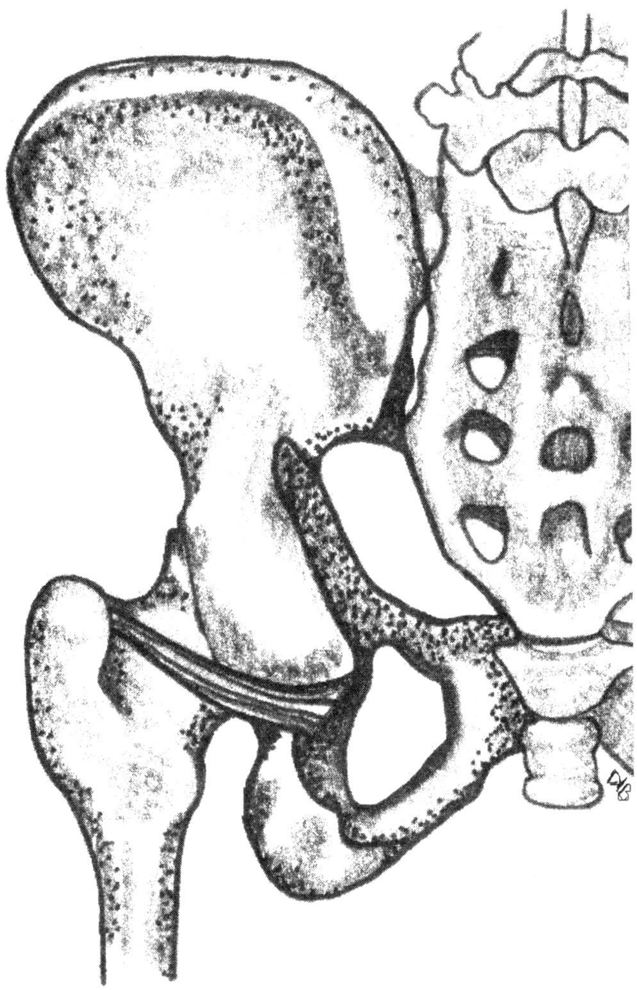

Origin:
Dorsal surface of Ischial spine (below the groove for the obturator internus tendon)

Insertion:
Medial surface of the greater trochanter (along with the tendon of obturator internus)

Action:
External rotation of femur (in the non-weight bearing, extended limb) Some authors also state that the gemelli may assist with abduction of the femur (with hip flexed at 90 degrees)

Innervation:
Quadratus femoris nerve (from lumbar & sacral nerves L-5 & S-1)

Relationship:
Both gemelli may be considered as accessory muscles to obturator internus. This muscle may be absent in some specimens. The gemelli and obturator muscles are collectively referred to as the **GOGO** muscles. Tenderness deep to the gluteals and inferior to the piriformis may be attributed to the gemelli.

OBTURATOR INTERNUS
Posterior View

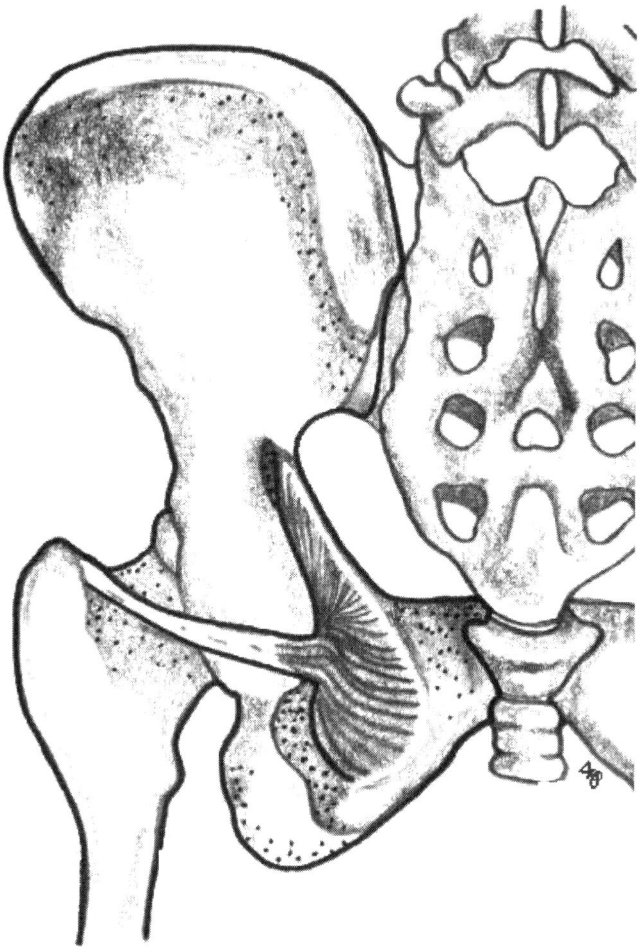

Origin:
Inner surface of the obturator membrane, the anterio-lateral wall of the lesser pelvic cavity and to a small extent, from the rim of the obturator foramen

Insertion:
Anterior portion of the medial surface of the greater trochanter (along with the tendon of the gemelli)

Action:
External rotation of femur (in the non-weight bearing, extended limb) Some authors also state that the obturators may assist with abduction of the femur (with hip flexed at 90 degrees) and stabilizing the hip during ambulation

Innervation:
Obturator Nerve (from spinal nerve branches L-5 & S-1)

Relationship:
Obturator internus lies partly within the true pelvis and partly posterior to the hip joint. The muscle exits the pelvis via the lesser sciatic foramen. A subtendinous bursa lies between the tendon of obturator internus and the capsule of the hip joint. Posterio-laterally, this muscle is covered by gluteus maximus and is passed by the sciatic nerve. The gemelli and obturators are collectively referred to as the **GOGO** muscles. Tenderness deep to the gluteals and inferior to the piriformis may be attributed to obturator internus.

OBTURATOR EXTERNUS
Posterior View

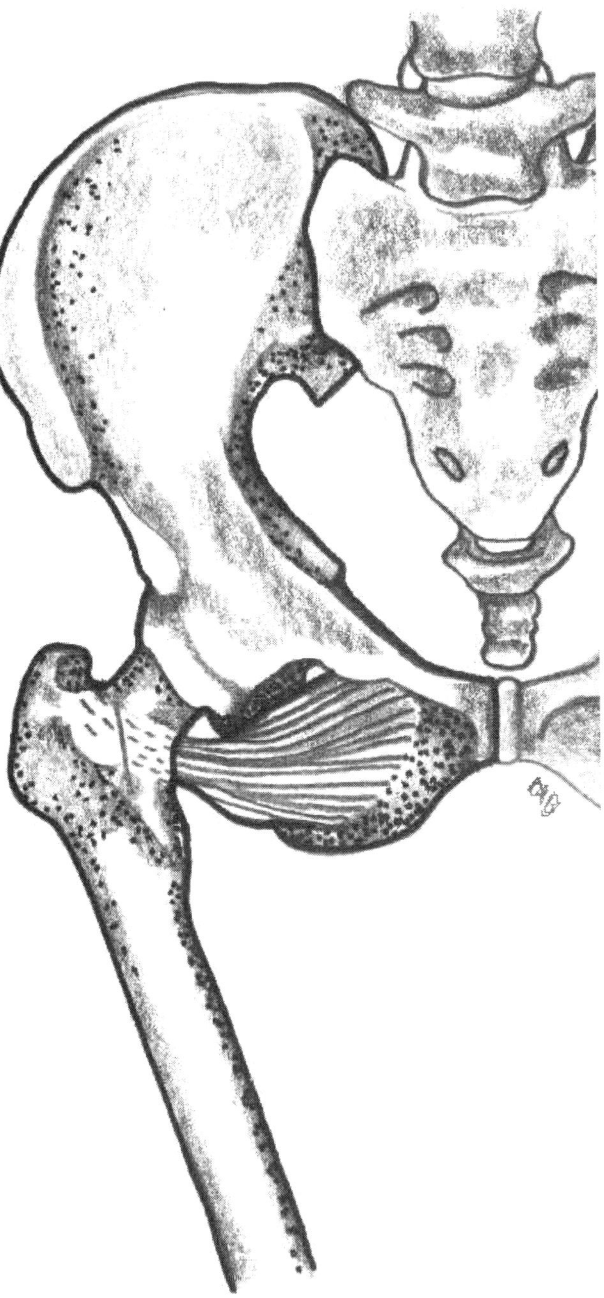

Origin:
Medial 2/3rd of the external surface of the obturator membrane and the adjacent portions of the pubic and ischial rami

Insertion:
Trochanteric fossa deep to the attachment of quadratus femoris

Action:
External rotation of femur (in the non-weight bearing, extended limb) Some authors also state that the obturators may assist with abduction of the femur (with hip flexed at 90 degrees) and stabilizing the hip during ambulation

Innervation:
Obturator Nerve (from spinal nerve branches L-3 & L-4)

Relationship:
Obturator externus is a flat, triangular muscle covering the exterior surface of the anterior pelvic wall. The tendon of obturator externus passes posterior to the hip joint capsule. A subtendinous bursa may be present. The gemelli & obturators are collectively referred to as the **GOGO** muscles. Tenderness deep to the gluteals and medial to the greater trochanter may be attributed to obturator externus.

QUADRATUS FEMORIS
Posterior View

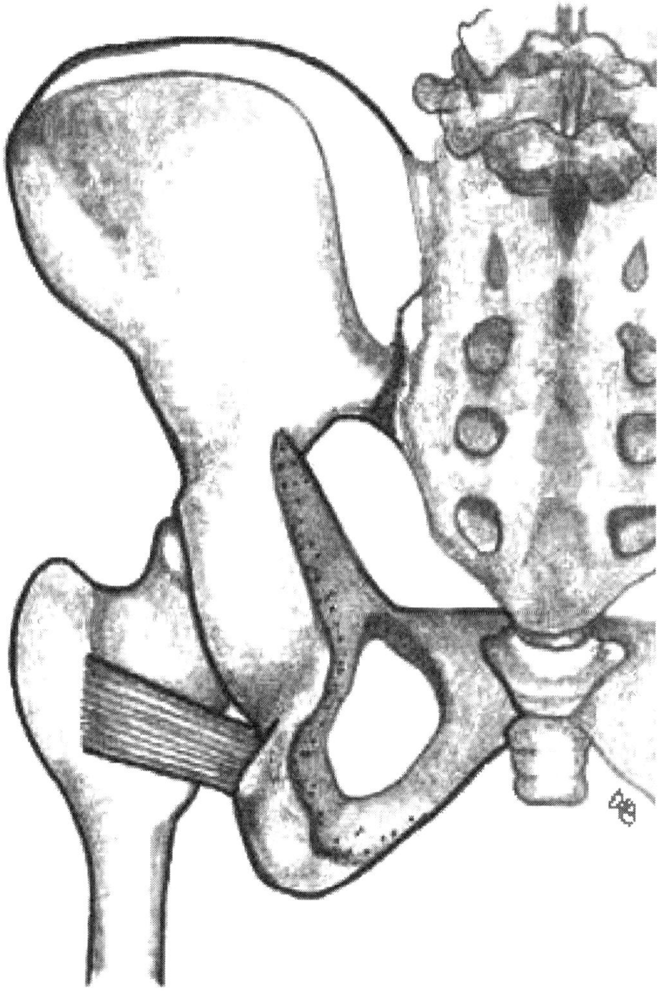

Origin:
Anterio-lateral surface of the ischium (superior aspect of the external portion of the ischial tuberosity)

Insertion:
The quadrate tubercle of femur and the intertraochanteric crest (extending about halfway between the greater & lesser tubercle)

Action:
External rotation of femur (in the non-weight bearing limb)
Stabilization of the hip during ambulation

Innervation:
Quadratus Femoris Nerve (from spinal nerve branches L-5 & S-1)

Relationship:
Quadratus femoris is a flat, quadrilateral muscle lying between gemellus inferior and the upper margin of adductor magnus. It may be palpated between the ischial tuberosity and the greater trochanter. A bursa is often present between the muscle and the lesser trochanter. This muscle may be absent in some specimens. Tenderness deep to the gluteals and medial to the greater trochanter may be attributed to quadratus femoris.

BICEPS FEMORIS (left)
Posterior View

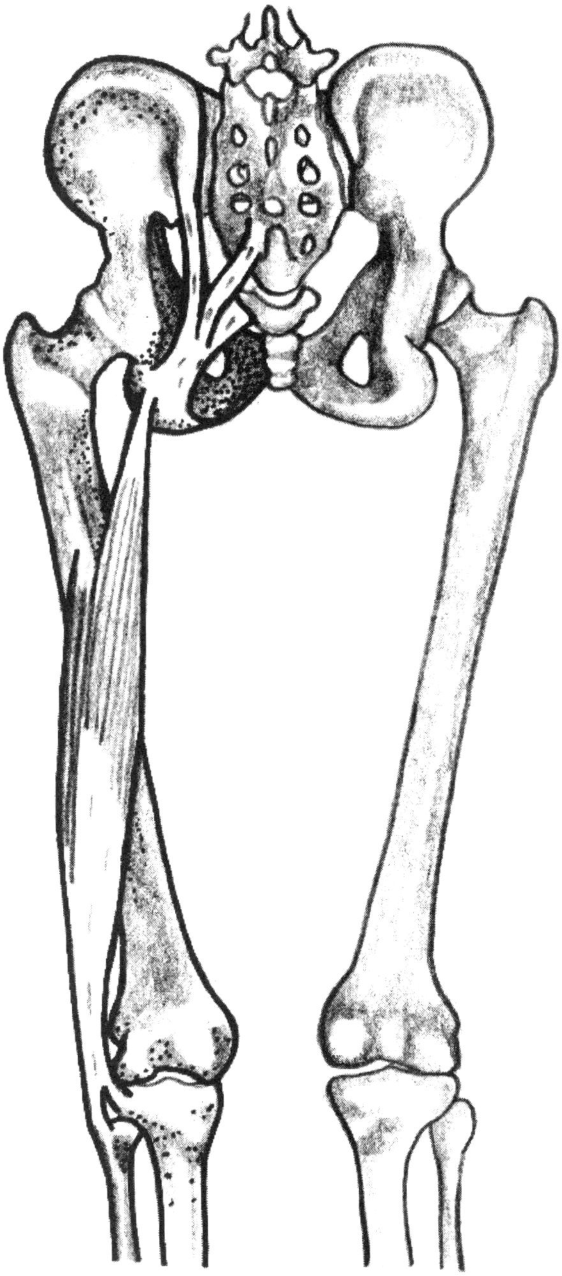

Origin:
Long head: Ischial tuberosity (inferior & posterior aspect via common tendon) and lower portion of the sacro-tuberous ligament
Short head: Linea aspera, along almost the entire length of the lateral lip

Insertion:
Fibula, lateral aspect of the head and lateral condyle of tibia (via an aponeurosis & 3 part tendon)

Action:
Extension of femur at hip, flexion of knee (only the short head is a pure knee flexor), assists in external rotation of the extended leg
Acting from below, it extends the hip from the flexed position

Innervation:
Long head: Sciatic nerve (tibial portion)
(via spinal nerves S-1, S-2, & S-3)
Short head: Sciatic nerve (common peroneal portion)
(via spinal nerves L-5, S-1, & S-2)

Relationship:
This most lateral hamstring muscle is a two-head muscle on the posterior lateral thigh. The long head crosses two-joints. The short head crosses only the knee and is sometimes absent. The muscle is active in walking, running, jumping, and bicycling. Hamstring curls (knee externally rotated) with resistance are useful exercises to strengthen this muscle. Hypertonic hamstrings are common in patients with low back pain. A limp often accompanies pain associated with this muscle group. Pain is often described as "sciatica-like" and must be distinguished from that of the deep rotators, spinal nerve-root origin, and osteoarthritis of the knee. Trigger point pain may be referred from the ischial tuberosity down the posterior aspect of the leg to mid calf.

SEMITENDINOSUS (left)
Posterior View

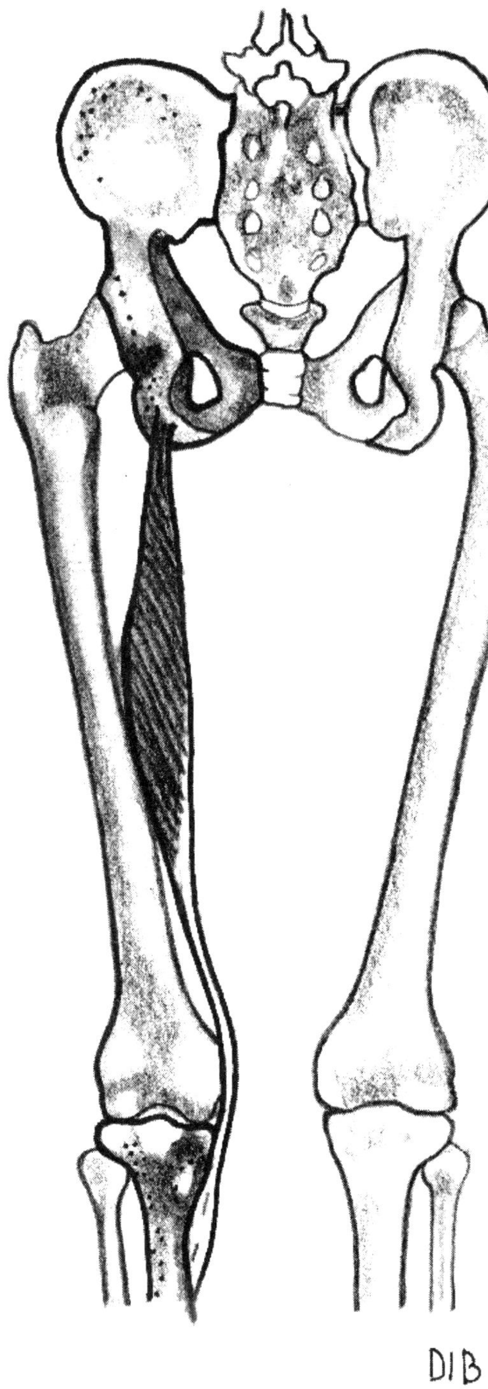

Origin:
Ischial tuberosity, (inferiomedial aspect) via a tendon shared with the long head of biceps femoris

Insertion:
Tibia (proximal portion, medial surface)
It is the most distal of the 3 tendons contributing to the pes anserinus

Action:
Extension of femur at hip, flexion of knee, assists in medial rotation of the extended leg. Acting from below, it extends the hip from the flexed position

Innervation:
Sciatic nerve (tibial portion) (via spinal nerves L-5, S-1, & S-2)

Relationship:
This muscle, on the posteromedial thigh, is known for its long, round tendon, which extends from midthigh to the tibia. The semitendinosus unites with the tendons of the sartorius and gracilis to form a flattened aponeurosis called the pes anserinus. Hamstring curls (knee internally rotated) with resistance are useful exercises to strengthen this muscle. Pain originating in semitendinosus presents like that of biceps femoris and must also be distinguished from previously noted sources. Quadriceps pain and gluteal hypotonicity may often be related to hamstring hypertonicity. The "straight-leg - raising" test will elicit pain and demonstrate limited motion with hamstring involvement. Trigger point pain may be referred down the posterior aspect of the leg to mid calf.

SEMIMEMBRANOSUS (left)
Posterior View

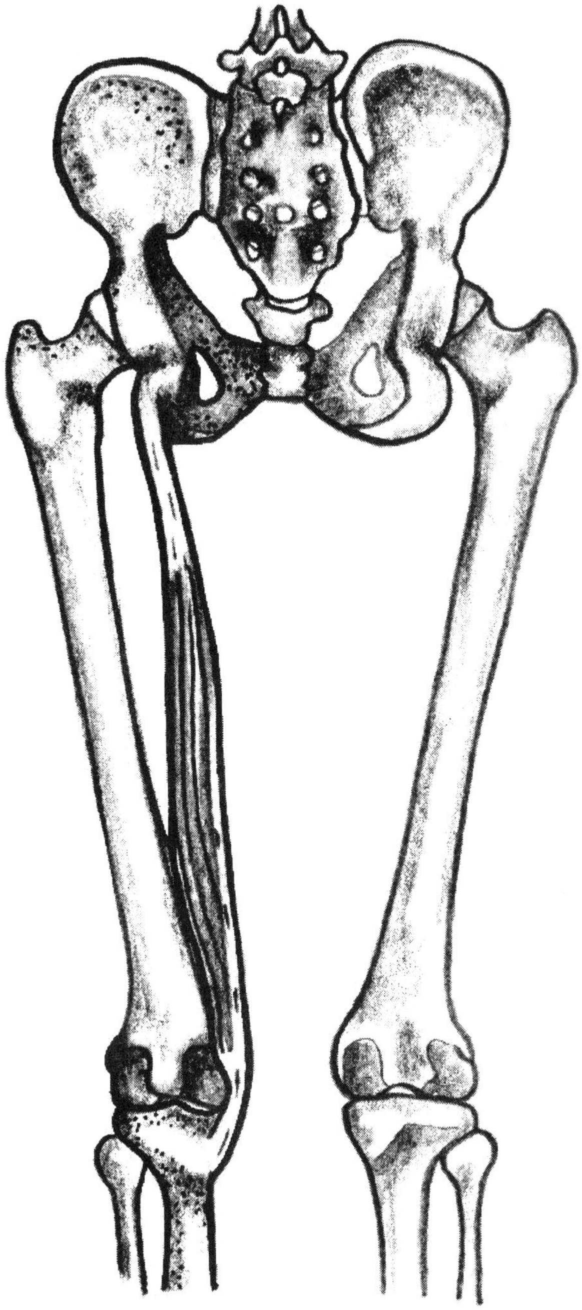

Origin:
Ischial tuberosity, superior and lateral aspect, and via a complex proximal shared tendon and aponeurosis

Insertion:
Tibia (posterior aspect of medial tibial condyle), Also, to the medial tibia and femoral condyle via fibrous expansion (which forms part of oblique popliteal ligament)

Action:
Extension of femur at hip, flexion of knee, assists in medial rotation of the extended leg. Acting from below, it extends the hip from the flexed position

Innervation:
Sciatic nerve (tibial portion) (via spinal nerves L-5, S-1, & S-2)

Relationship:
One of two medial hamstrings, the semimembranosus derives its name from its flat, membranous tendon of origin. Its fibers descend, sometimes intermingling with those of the semitendinosus and the biceps femoris. The semitendinosus overlaps this muscle throughout its extent. Semimembranosus functions with the other hamstrings as a myotatic unit in accomplishing their shared actions and as antagonists to the hip-flexors and extensors of the knee. Hamstring curls (knee internally rotated) with resistance are useful exercises to strengthen this muscle. Semimembranosus trigger point pain may be referred down the posterior aspect of the leg to mid calf.

GASTROCNEMIUS (left)
Posterior View

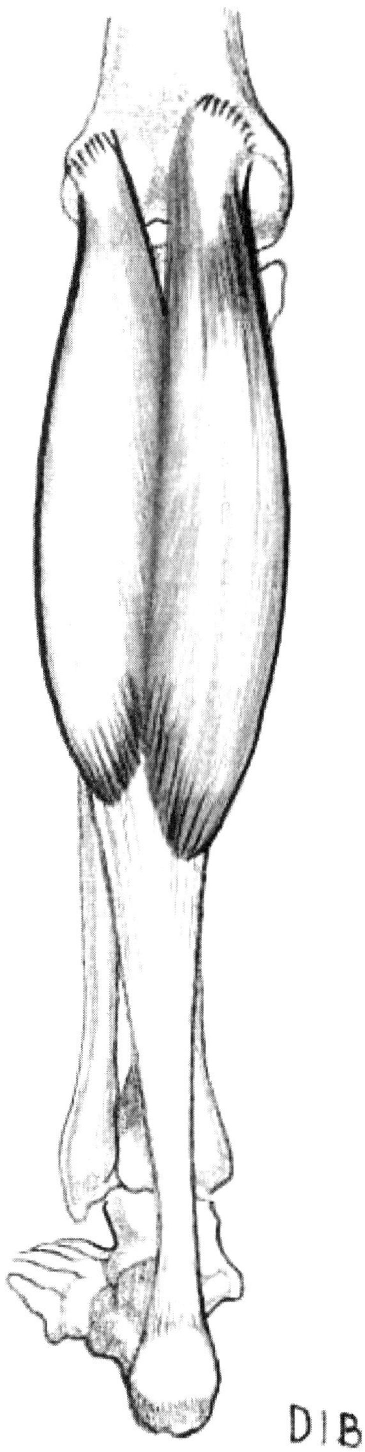

Origin:
Medial head: medial condyle of femur (upper/posterior part & popliteal surface adjacent to the condyle)
Lateral head: lateral condyle of femur (posterior/ lateral surface of condyle and supracondylar line)
Both heads: capsule of knee joint

Insertion:
Calcaneus, middle posterior surface, via the tendo calcaneus (A.K.A. Achilles Tendon)

Action:
Knee flexion, plantar flexion of foot (primarily when the knee is extended) and supination of the foot

Innervation:
Tibial nerve (via spinal nerves S-1 & S-2)

Relationship:
Gastrocnemius is the most superficial of the calf muscles, giving the characteristic contour to the calf. The medial head is the larger and its fibers extend further distally before spreading into a tendinous expansion where it is joined by the lateral head. The two heads join as the aponeurosis narrows, and with the tendon of soleus, form the tendo calcaneus, the thickest and strongest tendon in the human body. Gastrocnemius is active in walking, running, stair climbing, jumping, and bicycling. It contains a large number of slow-twitch, fatigue resistant, muscle fibers. Standing heel-raising with resistance is a useful exercise to strengthen this muscle. It is often involved in nocturnal calf cramps and intermittent claudication. Symptoms are aggravated by prolonged standing or walking on a slant, climbing a steep slope, or riding a bicycle with the seat adjusted too low. Gastrocnemius pain must be differentiated from that arising from the posterior knee, lumbar disc disease, phlebitis, and (plantar) foot dysfunction. Trigger point pain may be referred down the posterior leg to the sole and toes.

SOLEUS (left)
Posterior View

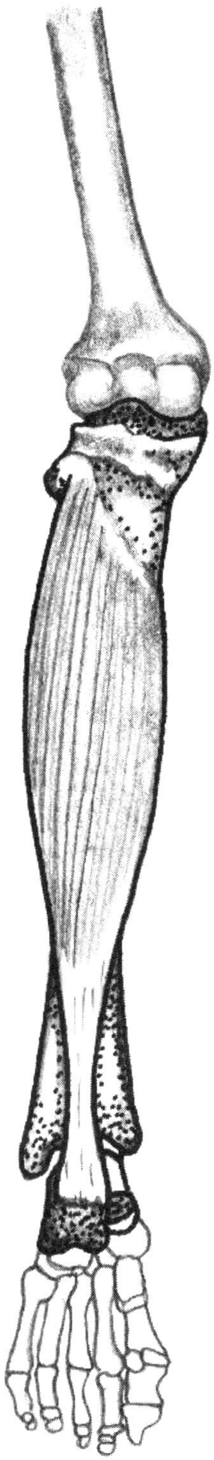

Origin:
Fibula (posterior surface of head & proximal 1/3 of shaft on posterior surface), Tibia (popliteal line and middle 1/3 of medial side of shaft) and arch of transverse intermuscular septum

Insertion:
Calcaneus, middle posterior surface, via a shared aponeurosis, which becomes the tendo calcaneus (A.K.A. Achilles Tendon)

Action:
Plantar flexion, assists in supination and is constantly active in quiet stance (it responds to the forward center of mass to prevent the body from falling forward)

Innervation:
Tibial nerve (via spinal nerves S-1 & S-2)

Relationship:
Soleus is a one-joint muscle, the largest of the "Triceps Surae". It is broad, flat, and lies just deep to the gastrocnemius. Its anterior attachment is a wide aponeurosis and most of its fibers course obliquely to the descending tendon on its posterior side. Soleus is active in walking, running, stair climbing, jumping, bicycling, and functions as a pump to assist venous blood flow from the feet and legs. Seated heel-raising with resistance is a useful exercise to strengthen this muscle. Soleus may be implicated in low back pain, heel pain, and "growing pains" of children. Symptoms are often aggravated by walking, running, skiing, skating, and by sustained contraction (as in wearing high heels). Soleus pain must be differentiated from that arising from disc disease (L-5/S-1), thrombophlebitis, Achilles tendinitis, stress fractures, shin splints, and systemic viral infection. Trigger point pain may be referred down the entire posterior calf to the sole and toes.

PLANTARIS (left)
Posterior View

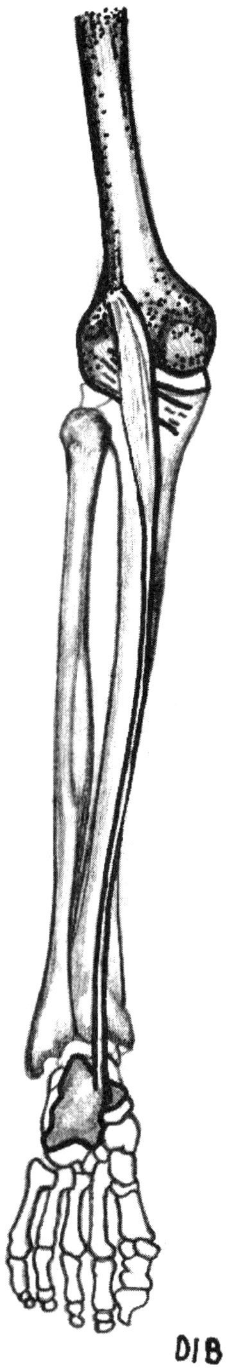

Origin:
Femur, lower portion of the lateral supracondylar line and the oblique popliteal ligament

Insertion:
Calcaneus (posterior/medial border)

Action:
Plantar flexion and assists in knee flexion

Innervation:
Tibial nerve (via spinal nerves S-1 & S-2)

Relationship:
Plantaris is a small, fusiform muscle lying between the gastrocnemius and the soleus. It is sometimes absent, at other times it is doubled. Its short belly is followed by a long slender tendon, which joins the medial border of the tendo calcaneus to insert on the posterior calcaneus. Plantaris is the lower extremity's equivalent of the palmaris longus and in some mammals, inserts directly into the plantar aponeurosis. In humans, it appears to be almost vestigial and plays a much less active roll.
Plantaris is most likely to be damaged by a sudden, forceful, lengthening contraction, and presents immediately with acute, sharp pain and later with ecchymosis of the lower, posterior leg.
Plantaris symptoms must be differentiated from those of thrombophlebitis, popliteal cysts, Achilles tendonitis, heel spurs, and gastrocnemius strain.
Trigger point pain of plantaris may be referred to the posterior knee and into the calf.

POPLITEUS (left)
Posterior View

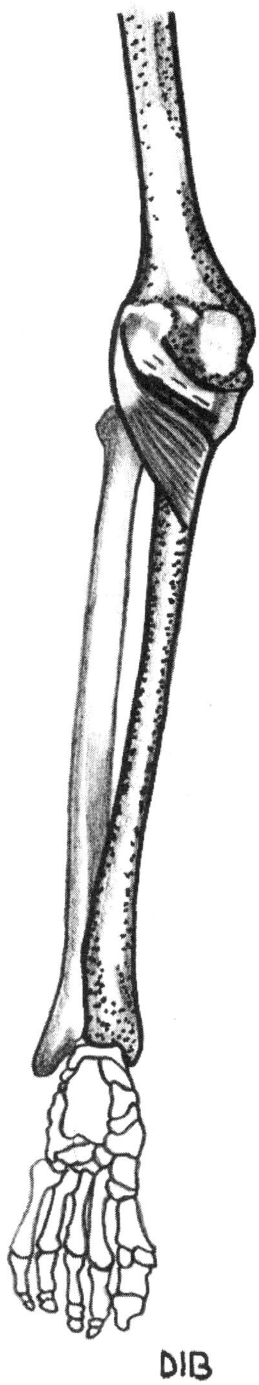

Origin:
Lateral condyle of femur (groove on anterior surface), Arcuate popliteal ligament and a portion of the capsule of knee adjacent to lateral meniscus

Insertion:
Tibia (posterior triangular surface above the soleal line)

Action:
Flexion of knee (from fully extended position)
Internal rotation of tibia (when proximal attachment is fixed)
External rotation of femur (when distal attachment is fixed)

Innervation:
Tibial nerve (via spinal nerves L-4, L-5, & S-1)

Relationship:
Poplitius is a flat muscle, which arises within the capsule of the knee and forms the lower floor of the popliteal fossa. It is said to "unlock" the knee from the fully extended position and may assist in retracting the posterior meniscus during flexion of the knee with lateral rotation of the femur. Poplitius is very active in crouching, thus assisting the posterior cruciate ligament in preventing forward dislocation of the femur upon the tibia. Isolated internal rotation and flexion of the knee against resistance strengthen poplitius. Complaints from poplitius usually arise from overuse or trauma in crouching, running, and walking. Symptoms are made worse by ambulating downhill or down stairs. Poplitius symptoms must be differentiated from those of popliteal cysts, deep thrombosis, knee instability, and meniscus (or capsular) tear. Trigger point pain of poplitius may be referred to the posterior knee.

TIBIALIS POSTERIOR (left)
Posterior View

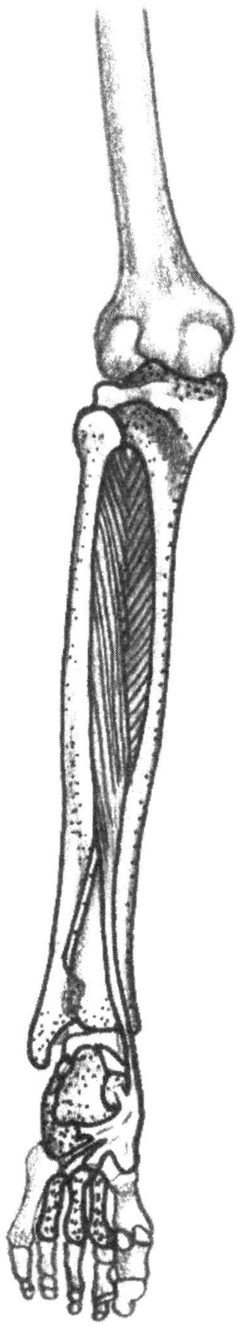

Origin:
Tibia: (proximal 2/3 of posterior shaft as well as the distal lateral condyle)
Fibula: (proximal 2/3 of shaft and posterior surface of head)
Interosseus membrane (along entire posterior surface)

Insertion:
Navicular (tuberosity)
Calcaneus (sustentaculum tali)
All three cuneiform bones and bases of the
2nd, 3rd, and 4th metatarsals

Action:
Inversion of foot may assist in (forceful) plantar flexion, Supination (inversion & adduction) during weight bearing.

Innervation:
Tibial nerve (via spinal nerves L-4 & L-5)

Relationship:
Most deeply placed of the flexor group, high on the posterior leg, tibialis posterior is overlapped by both the flexor hallucis longus and the flexor digitorum longus. It arises from two narrow heads and descends centrally on the leg to form its tendon distally. This tendon passes behind the medial malleolus (with the flexor digitorum longus), enters the foot on the plantar surface (where it contains a sesamoid bone), and then divides to its several insertions. During weight bearing, tibialis posterior assists in elevating the longitudinal arch of the foot and in limiting pronation. It is active in walking and running, especially on uneven surfaces. Heel-raising exercises with foot inverted are useful to strengthen this muscle. Weakness or absence results in excessive pronation or "pes valgus deformity". Symptoms usually include severe pain in the sole of the foot, Achilles tendon, and (to a lesser degree) the heel and calf. Tibialis posterior may be involved in shin splints and posterior compartment syndrome, and should be distinguished from other sources of ankle and foot pain. Trigger point pain of tibialis posterior may be referred to the knee and down the posterior calf.

FLEXOR DIGITORUM LONGUS (left)
Posterior View

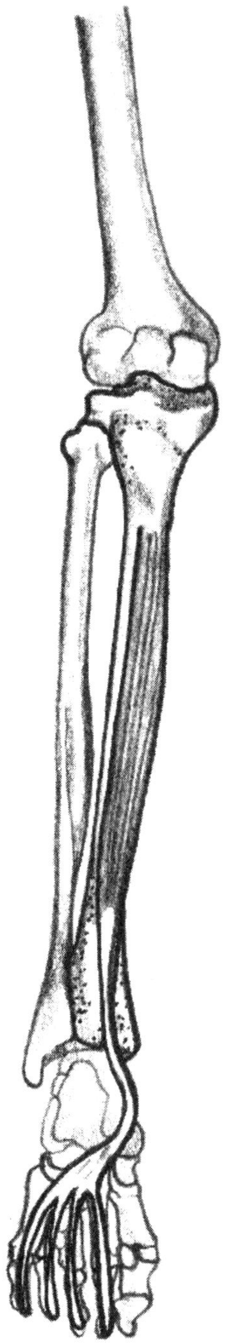

DJB

Origin:
Tibia (posterior surface of middle 2/3 of shaft) and fascia covering the tibialis posterior

Insertion:
Distal phalanges (digits 2 to 5 at base)

Action:
Flexion of distal phalanges (digits 2 to 5), assists in plantar flexion & inversion of foot

Innervation:
Tibial nerve (via spinal nerves L-5, S-1, & S-2)

Relationship:
Flexor digitorum longus lies deep on the tibial side of the leg. It is thin proximally, but increasingly widens as it descends. The muscle ends in a tendon, which extends along nearly the entire posterior surface of tibialis posterior, eventually passing posteriorly to the medial malleolus and being joined in the sole of the foot by the tendon of the quadratus plantae. The tendon then divides distally into four slips, which insert into the four lateral toes. Walking barefoot on soft sand, gripping with the toes, and forced inversion, exercise this muscle. Grabbing a towel with the toes and pulling medially against resistance, with the heel planted, specifically strengthens it. Flexor digitorum longus is active in standing, walking, running, and jumping. Complaints of toe and plantar foot pain while walking and medial ankle pain are common with involvement of flexor digitorum longus. These symptoms should be differentiated from those of "tarsal tunnel syndrome" and cramping of the intrinsic flexors of the toes. "hammer toe" and "clawtoe" deformities may also be associated with dysfunction of this muscle. Trigger point pain may be referred down the posterior calf into the heel.

FLEXOR HALLUCIS LONGUS (left)
Posterior View

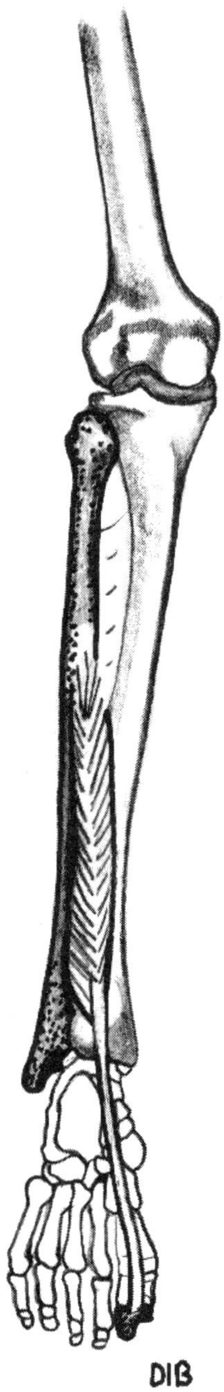

Origin:
Fibula (distal 2/3 of posterior surface), fascia of tibialis posterior and from the interosseous membrane

Insertion:
Hallux (plantar aspect of base of distal phalanx)

Action:
Flexion of distal phalange of Hallux (large toe), assists in plantar flexion & inversion of foot

Innervation:
Tibial nerve (via spinal nerves L-5, S-1, & S-2)

Relationship:
Flexor hallucis longus is a thin muscle lying distal and lateral to flexor digitorum longus and tibialis posterior, deep to soleus and gastrocnemius, and medial to the peronei. In the sole of the foot, the tendon passes between the two tendons of flexor hallucis brevis to its insertion at the base of the great (large) toe. It is most active in powerful flexion of the distal phalanx of the great toe, while only weakly flexing the proximal phalanx. This action is most demonstrated in maintaining balance while standing on the toes. During ambulation, this muscle is used to stabilize the foot and ankle during take-offs and landings. Towel grabs (as described with flexor digitorum longus) are used to specifically strengthen it. Complaints involving flexor hallucis longus are similar to those of flexor digitorum longus. Hallux valgus (bunion) is commonly associated with dysfunction of this muscle. Trigger point pain of flexor hallucis longus may be referred down the posterior calf into the heel and sole.

STERNOCLEIDOMASTOID (left)
Lateral View

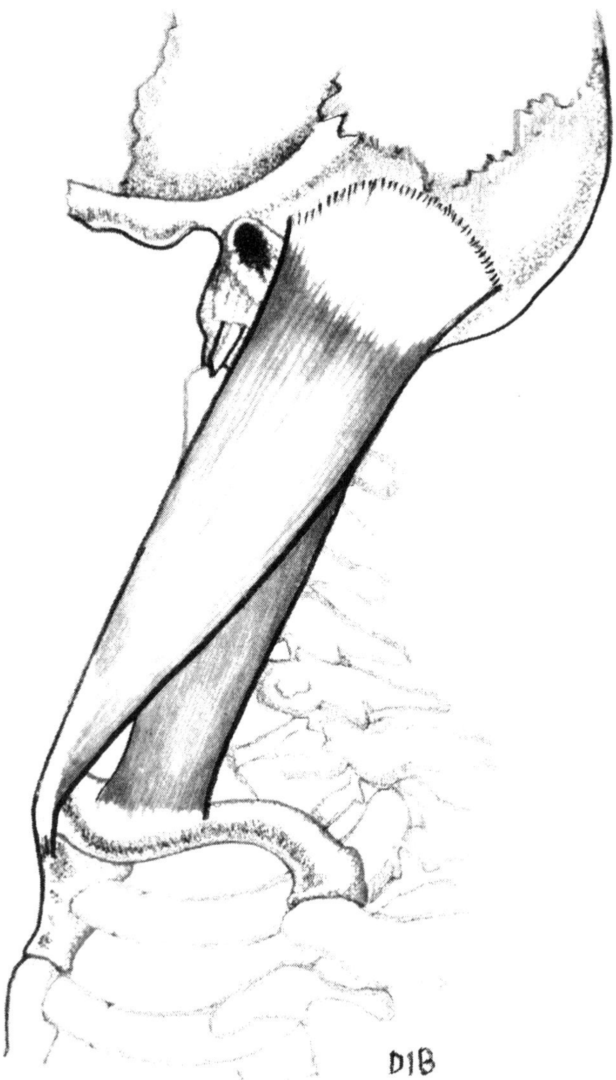

Origin:
Sternal (medial) head: Sternum and Manubrium (ventral surface)
Clavicular (lateral) head: Clavicle (superior and anterior surface of medial 1/3)

Insertion:
Mastoid process of temporal bone (lateral surface), Occiput (lateral half of superior nuchal line)

Action:
Bilaterally: Flexion of cervical spine, capital extension (posterior fibers), checkrein hyperextension of head/neck and raise sternum in forced inspiration
Unilaterally: Rotation of head to opposite side and side-bending of cervical spine to same side

Innervation:
Accessory Nerve (Cranial Nerve XI) (via spinal nerves C-2 & C-3, ventral rami)

Relationship:
The two heads of the SCM gradually merge in the neck as the muscle rises upward laterally and posteriorly. The sternal head is more superficial and more oblique. Each SCM may act as a synergist with upper trapezius and scalenes of the same side. Flexion of the cervical spine against resistance is a useful exercise to strengthen this muscle. The SCM are frequently aggravated by overuse (working overhead, etc.), sports injury, chronic cough, or whiplash. Common SCM related symptoms include frontal headache, postural vertigo, facial neuralgia, and torticolis (wry-neck). Trigger points refer pain to the vertex, occiput, eye and face (sternal head), and forehead and ear (clavicular head). These complaints should be differentiated from those of vascular and neurological origin.

SCALENES (right)
Anterolateral View

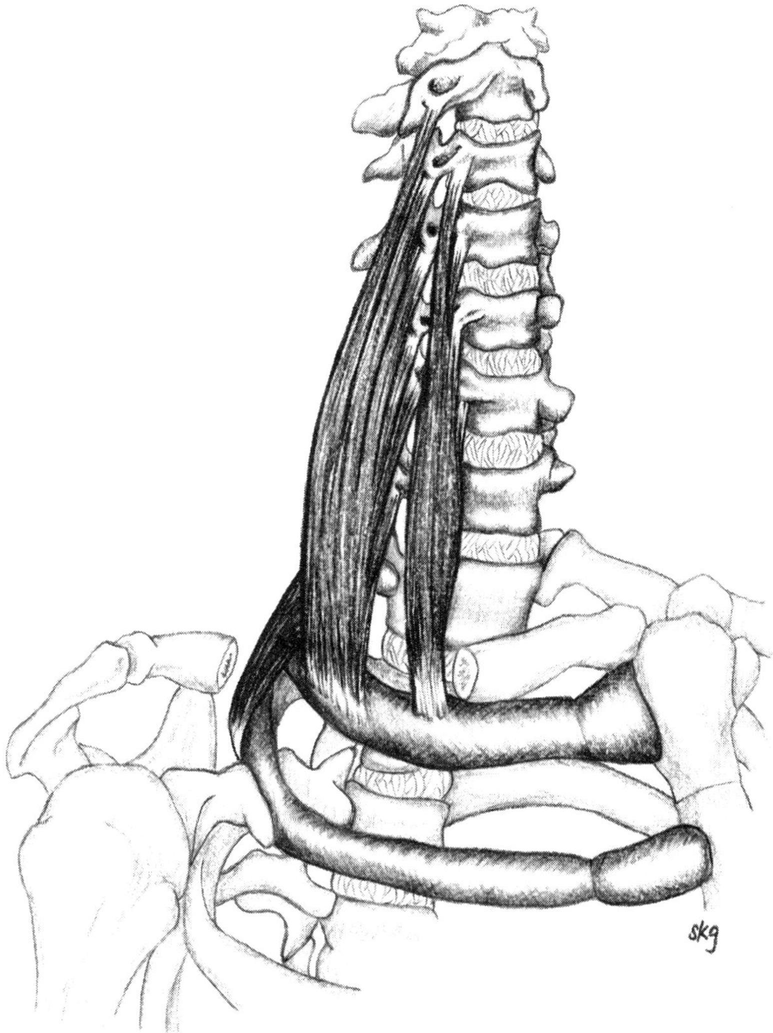

Origin:
Anterior: C-3 thru C-6 vertebrae (anterior tubercles of transverse processes)
Middle: C-2 thru C-7 (posterior tubercles of transverse processes) and axis
Posterior: C-4 thru C-7 vertebrae (posterior tubercles of transverse processes)

Insertion:
Anterior: 1st rib (superior & medial surface)
Middle: 1st rib (superior surface)
Posterior: 2nd rib (lateral surface)

Action:
Bilaterally: Flexion of cervical spine and elevation of 1st & 2nd ribs in inspiration
Unilaterally: side-bending of neck to same side and assists slightly in rotation of cervical spine to opposite side

Innervation:
Spinal nerves C-3 thru C-8

Relationship:
All three of the scalenes are located deep to the sternocleidomastoid muscle and their attachments are highly variable. The middle is the longest and largest of the scalenes. The posterior is the smallest and deepest and is often fused with scalenus medius except when scalenus minimus is present (50% to 75% of specimens). The brachial plexus and subclavian artery pass between scalenus anterior and medius and may occasionally be entrapped. The scalenes may be aggravated by trauma, overuse (in pulling, swimming, carrying), scoliosis, chronic cough, or poor sleeping posture. Symptoms include local pain, referred pain, and neurological disturbances. Trigger points refer pain to the upper vertebral border of the scapula, pectoral region, down the dorsal and ventral surfaces of the arm (skipping the elbow) to the radial forearm, and into the thumb and index finger. These complaints must be differentiated from those of carpal tunnel syndrome, angina, cervical joint or disc disease, and the possibility of a cervical rib.

SUBCLAVIUS (left)
Anterior View

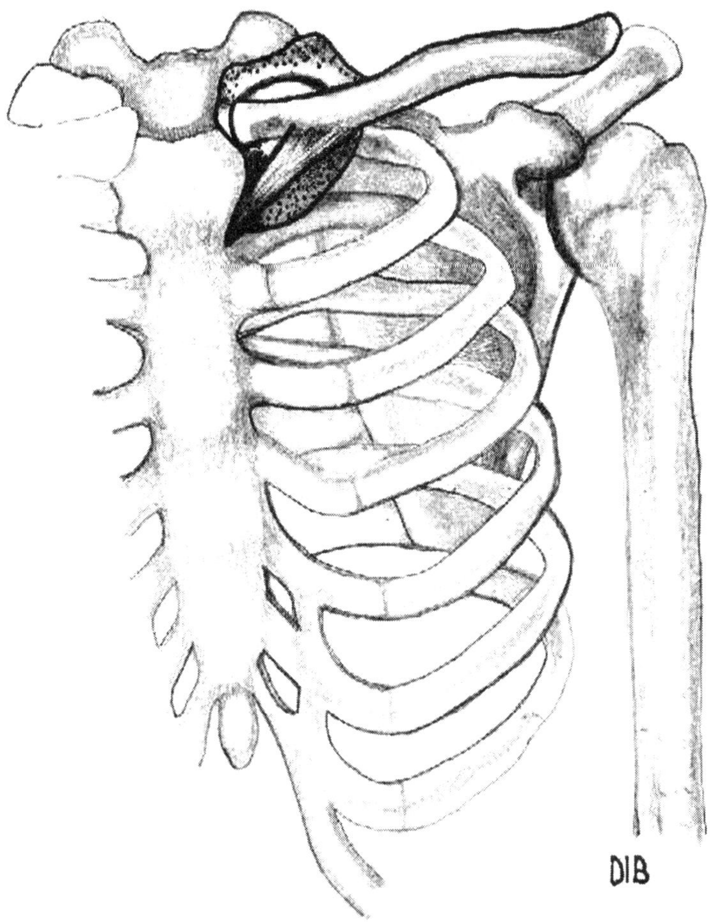

Origin:
Rib 1 and its cartilage (at their junction)

Insertion:
Clavicle (inferior surface, groove in medial 1/3)

Action:
Shoulder depression (assists),
Depresses and moves clavicle anteriorly, thus stabilizing it during shoulder motion

Innervation:
Spinal Nerves C-5 & C-6 (subclavian branch via brachial plexus)

Relationship:
The subclavius is a small, elongated muscle lying inferior to the clavicle (between it and the 1st rib). Its fibers run upward and laterally following the contour of the clavicle. Shortening of the subclavius muscle can contribute to symptoms of thoracic outlet syndrome by depressing the clavicle and thus compressing neurovascular components as they pass superficial to the 1st rib. Subclavius may also refer pain across the anterior shoulder and distally along the anterior and radial portion of the arm and forearm. Symptoms should be differentiated from angina pectoris, biceps tendonitis, subacromial bursitis, and cervical joint/disc disease.

PECTORALIS MAJOR (left)
Anterior View

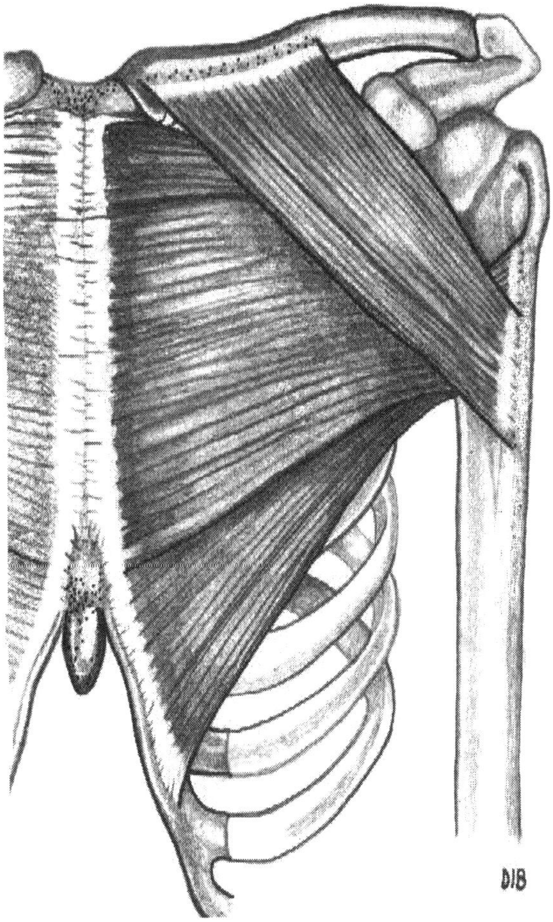

Origin:
Clavicular portion: Clavicle (anterior surface of sternal half)
Sternocostal portion: Sternum, cartilage of 2nd thru 6th ribs and the abdominal aponeurosis

Insertion:
Humerus (lateral lip of bicipital groove)

Action:
Adduction & medial (internal) rotation of humerus, flexion & horizontal adduction of humerus and elevation of thorax in forced inspiration (with both upper extremities fixed)

Innervation:
Medial & Lateral Pectoral nerves (via spinal nerves C-5 thru T-1)

Description:
This muscle is a large, thick, fan-shaped muscle covering the anterior and superior surfaces of the thorax, and forming part of the anterior wall of the axilla. The muscle is divided into two major portions that converge toward the axilla. The clavicular fibers pass downward and laterally toward the humeral insertion. The sternocostal (deep) fibers pass horizontally from midsternum and upward and laterally from the rib attachments. The lower fibers rise almost vertically toward the axilla to join a common tendon of insertion on the humerus. This muscle is active in walking with crutches, swimming, golfing, and throwing a baseball. Bench presses are often used to strengthen pectoralis major. Complaints associated with pectoralis major include local and referred trigger point pain (down the ulnar aspect of the arm to the 4th and 5th digits), weakness, and reduced ROM. Symptoms are aggravated by prolonged sitting (reading, writing, etc.), heavy lifting, and immobilization. A round-shouldered posture may also be present. Complaints must be differentiated from those of cardiovascular origin, bursitis, tendonitis, epicondylitis, hiatal hernia, pulmonary disease, and cervical joint/disc disease.

PECTORALIS MINOR (left)
Anterior View

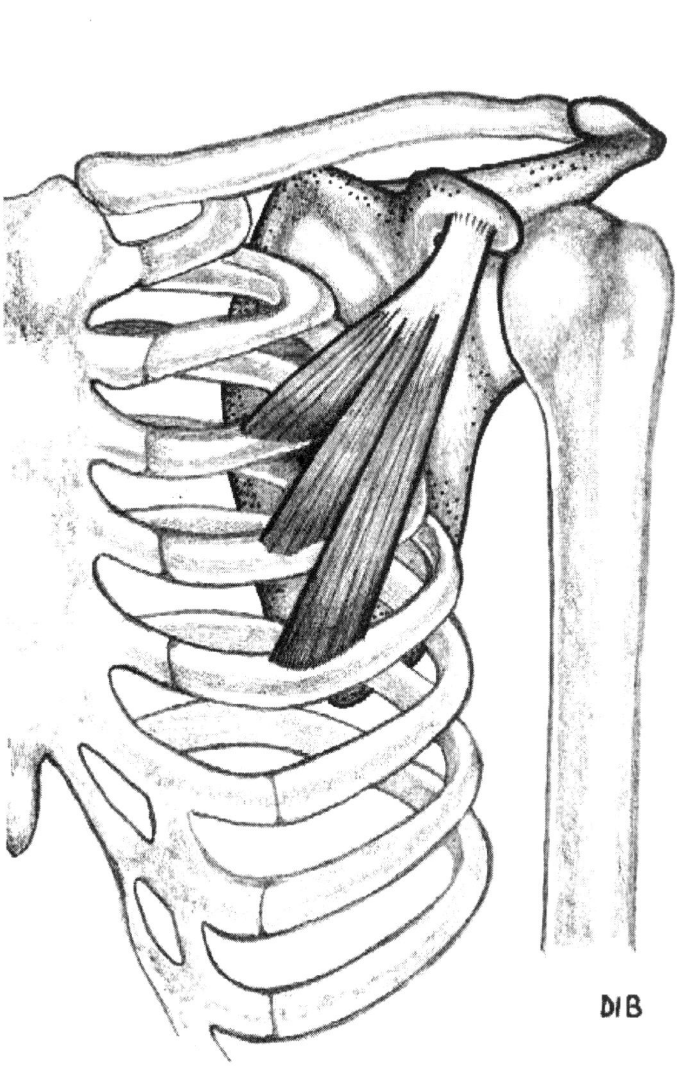

Origin:
Ribs 3 to 5 (near costal cartilages)
Aponeurosis of intercostals muscles

Insertion:
Coracoid Process of scapula (medial border and superior surface of tip)

Action:
Scapular protraction (forward, obliquely downward and medial movement)
Elevation of ribs in forced inspiration when the scapula is fixed by the levator scapulae

Innervation:
Medial pectoral nerve (via spinal nerves C-8 & T-1)

Relationship:
This muscle, broader at its origins, lies on the upper thorax directly deep to the pectoralis major. The fibers pass upward and laterally to converge into a flat tendon. It forms a minor portion of the anterior wall of the axilla. Pectoralis minor is active in crutch walking and forceful adduction and flexion of the humerus. Dips and corner push-ups are used to strengthen this muscle. It is aggravated by prolonged sitting (while reading, writing, etc.), poor posture, carrying a heavy shoulder bag, and by trauma. Complaints of local and referred pain are often accompanied by a round-shouldered posture and decreased ROM in reaching upward and backward. Trigger points refer pain to the chest and ulnar aspect of the arm. Compression of the axillary artery and brachial plexus is common. Symptoms must be differentiated from those of scalene entrapment, cervical joint/disc disease, angina, tendonitis, and epicondylitis.

DIAPHRAGM Anterior View

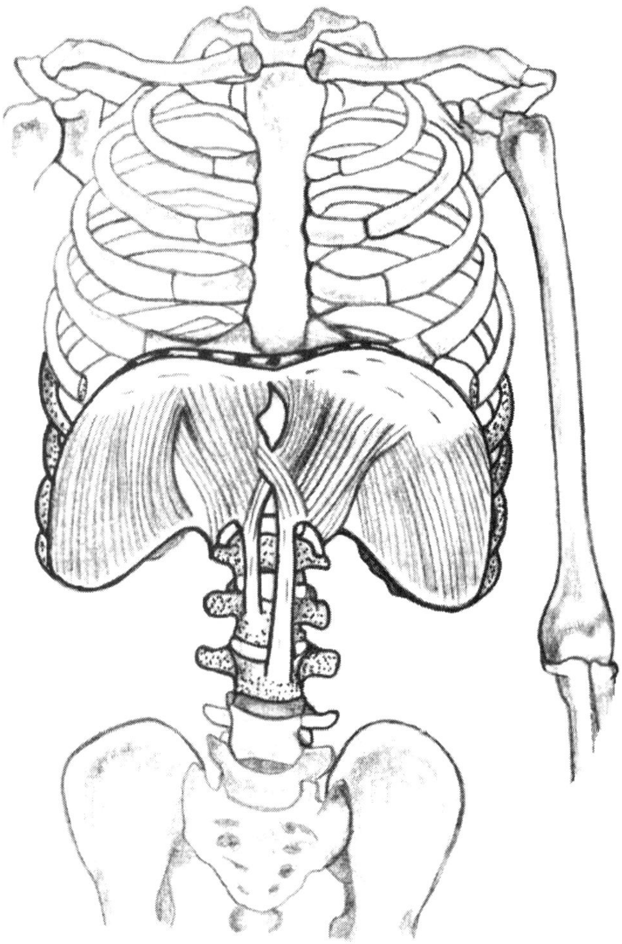

Origin:
Circumference of the thoracic outlet: Xiphoid process (posterior surface), Ribs 7 thru 12 (bilaterally, inner surfaces of the cartilage and ribs), Vertebrae: L-1, L-2, & L-3 (from the medial and lateral arcuate ligaments and from bodies of the vertebrae by two muscular crura)

Insertion:
Central Tendon of diaphragm

Innervation: Phrenic Nerve (via spinal nerves C-3 thru C-5)

Action:
Inspiration: Contraction of the diaphragm draws the central tendon downward and forward. This increases the vertical thoracic dimensions and pushes the abdominal viscera downward. It decreases the pressure within the thoracic cavity, allowing air to be forced into the lungs through the open glottis by the higher pressure of the atmospheric air. Expiration: Passive relaxation allows the half-dome to ascend, thus decreasing thoracic cavity volume and increasing its pressure

Relationship:
This half-dome shaped muscle of contractile and fibrous tissue forms the floor of the thorax and the roof of the abdomen. The diaphragm is muscular on the periphery and is tendinous in its central area. The muscle is flatter centrally than at the periphery and higher on the right (reaching rib 5) than on the left (reaching rib 6). From the peak on each side, the diaphragm abruptly descends to costal and vertebral attachments. Complaints associated with the diaphragm include chest pain, dyspnea, hiccup, and stitch in the side. Trigger points refer pain to the upper border of the ipsilateral shoulder or the costal area. These symptoms are aggravated by exercise, persistent cough, and high-anxiety situations. They should be differentiated from those of peptic ulcer, gastric reflux, and gall bladder disease.

RECTUS ABDOMINIS (left)

Anterior View

Origin:
Pubis (tubercle on crest)
Ligaments covering anterior of symphysis pubis

Insertion:
Ribs 5 to 7 (costal cartilages by three fascicles of differing size)
Sternum & Xiphoid (costo-xiphoid ligaments)

Action:
Flexion of spine (draws symphysis and sternum toward each other), with other abdominal muscles compresses abdominal contents.
Posterior tilt of pelvis

Innervation:
Intercostal Nerves T-7 thru T-12 (via spinal nerves T-7 thru T-12)

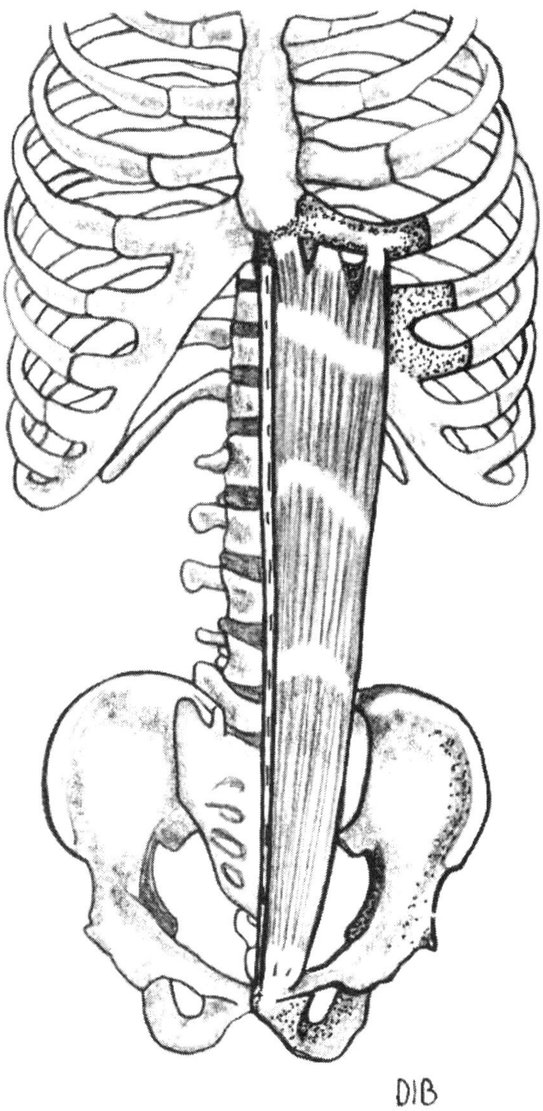

Relationship:
This is a long muscular strap extending from the ventral lower sternum to the pubis. Rectus is enclosed within a fascial sheath whose lateral borders form the abdominal aponeurosis. Its vertical fibers lie centrally along the abdomen, each separated from its contralateral partner by the linea alba. The muscle is interrupted (but not all the way through) by three (or more) fibrous bands called the tendinous inscriptions, which pass transversely across the muscle. The most superior band is generally at the level of the xiphoid; the lowest inscription is at the level of the umbilicus, and the second intersection is midway between the two. This muscle is active during walking, running, and jumping. Doing sit-ups or crunches with knees flexed engages rectus abdominis, while passively shortening iliopsoas and rectus femoris. Complaints associated with rectus abdominis include local and referred pain and visceral disorders. Symptoms are aggravated by internal disease, stress, trauma, and overuse. They should be differentiated from articular dysfunction, gastro-intestinal disease, and genito-urinary disorders.

PYRAMIDALIS
Anterior View

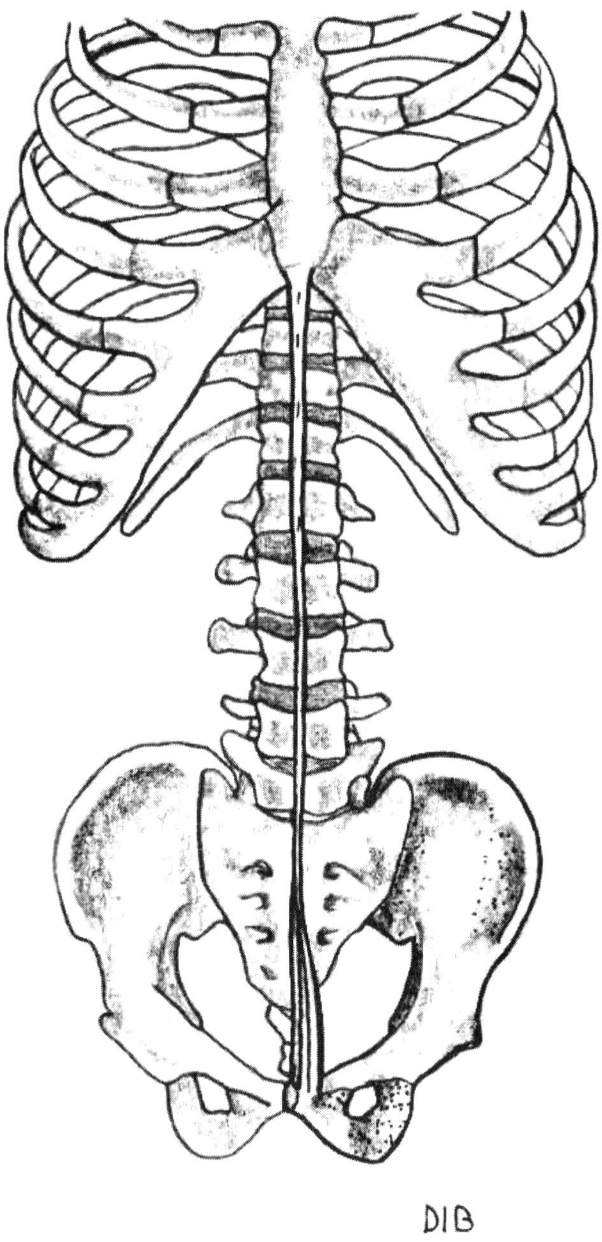

Origin:
Pubis (anterior surface of ramus)

Insertion:
Linea alba (midway between umbilicus and pubis)

Action:
Tenses the linea alba

Innervation:
Spinal nerve T-12 (ventral ramus)

Relationship:
Pyramidalis is a small triangular-shaped muscle located in the extreme distal portion of the abdominal wall and lying anterior to the lower rectus abdominis. It is found within the fascial sheath of rectus abdominis. Its origin on the pubis is wide and narrows as it rises to a pointed insertion. The muscle varies considerably from side to side and may be present or absent (in approximately 17% of specimens). Pain in the midline, between the umbilicus and symphysis pubis, may be associated with pyramidalis dysfunction.

EXTERNAL OBLIQUE (left)
Lateral View

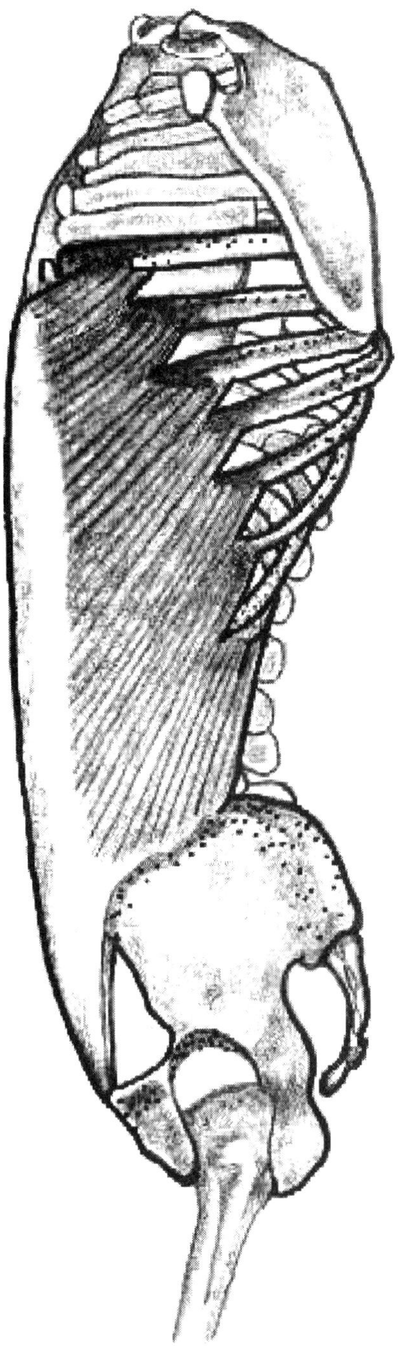

Origin:
Ribs 5 thru 12 (external surfaces & inferior borders by digitations that appear continuous with the external intercostals and interdigitate with serratus anterior and latissimus dorsi)

Insertion:
Abdominal aponeurosis (& linea alba) and anterior ½ of iliac crest

Innervation:
Intercostal Nerves T-8 thru T-12 (via spinal nerves T-7 thru T-12)

Action:
Bilaterally: Flexion of spine and increase abdominal pressure to assist in defecation and force viscera upward during expiration to elevate diaphragm
Unilaterally: Lateral flexion of spine to same side and rotation of trunk toward the contralateral side

Relationship:
External oblique is the most superficial of the muscles of the lateral abdominal wall. Its fibers run obliquely inferior and medially. External oblique is active during walking, running, and to checkrein forceful rotational movements of the trunk. This muscle is also involved in pumping blood and lymph upward through the abdomino-pelvic cavity. Twisting elbow to knee, when doing sit-ups or crunches with knees flexed, engages external oblique, while passively shortening iliopsoas and rectus femoris. External oblique may present with complaints of local and referred pain, nausea, emesis, diarrhea, and urinary dysfunction. Complaints are often initiated by prolonged overuse, trauma, visceral disease, and stress. Symptoms must be differentiated from those of hiatal hernia, peptic ulcers, colitis, appendicitis, cholecystitis, fibromyalgia, and articular dysfunction.

INTERNAL OBLIQUE (right)
Anterolateral View

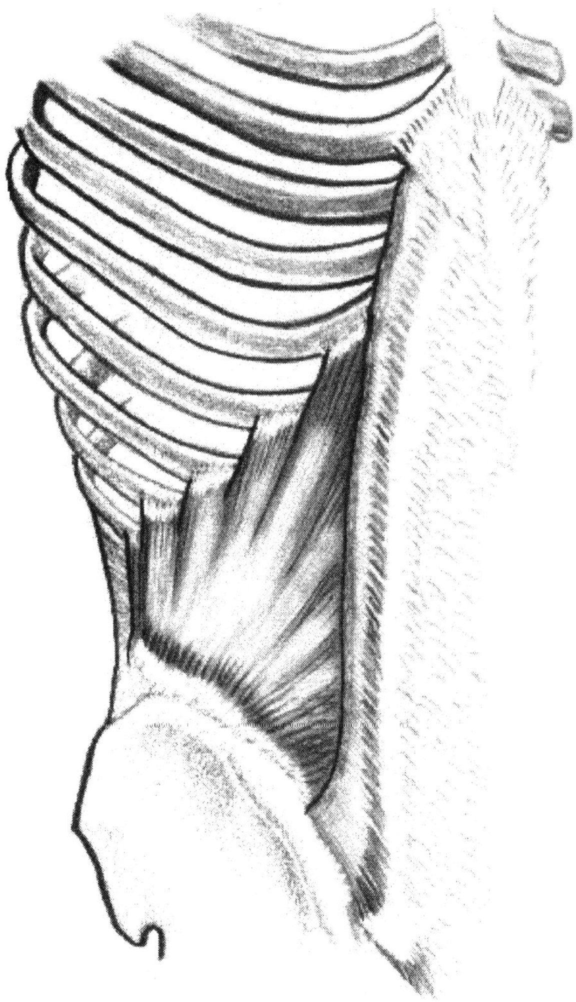

Origin:
Inguinal ligament (lateral 1/2)
Iliac crest (anterior 2/3)
Thoracolumbar fascia (lower portion)

Insertion:
Ribs 9 to 12 (inferior borders by digitations that appear continuous with internal intercostals), Cartilage of ribs 9 to 12 (via an aponeurosis), Aponeurosis that fuses with the external oblique and into the linea alba, Pubis (pectineal line, with tendinous sheath from transversus abdominis)

Action:
Bilaterally: Flexion of spine and increase abdominal pressure to assist in defecation and force viscera upward during expiration to elevate diaphragm
Unilaterally: Lateral flexion of spine to same side and rotation of trunk toward the same side

Innervation:
Intercostal Nerves T-8 thru T-12 (via spinal nerves T-8 thru L-1)

Relationship:
This muscle is smaller and thinner than the external oblique, under which it lies on the lateral and ventral abdominal wall. The fibers from the iliac crest pass upward and medially to ribs 9 to 12 and the aponeurosis (the more lateral the fibers lie, the more they run toward the vertical). The lowest fibers pass almost horizontally on the lower abdomen. Strengthening exercises given for the external oblique also apply to this muscle. Internal oblique may present with complaints of local and referred pain, nausea, emesis, diarrhea, and urinary dysfunction. Complaints are often initiated by prolonged overuse, trauma, visceral disease, and stress. Symptoms must be differentiated from those of hiatal hernia, peptic ulcer, colitis, appendicitis, cholecystitis, fibromyalgia, and articular dysfunction.

TRANSVERSE ABDOMINIS (left)
Lateral View

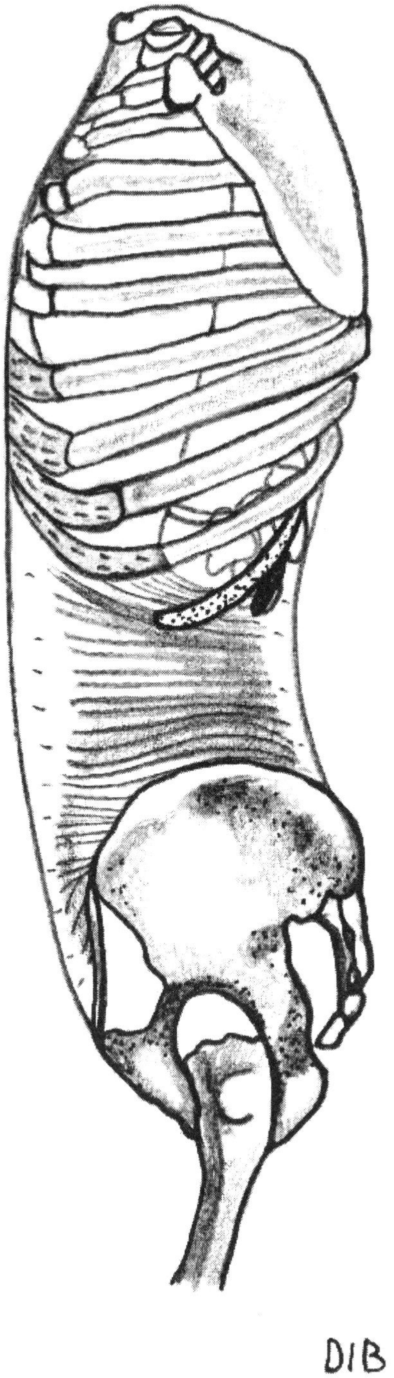

Origin:
Inguinal ligament (lateral 1/3)
Iliac crest (anterior 3/4)
Thoracolumbar fascia (between iliac crest and 12th rib)
Costal cartilages 7 thru 12 (inner surfaces)

Insertion:
Linea alba (via sheath of rectus abdominis),
Pubis (via the conjoint tendon/inguinal falx)

Innervation:
Intercostal Nerves T-7 thru T-12 (via spinal nerves T-8 thru L-1)

Action:
Constricts (flattens) abdomen, compressing the abdominal viscera and assisting in expelling their contents, forced expiration

Relationship:
The innermost of the flat abdominal muscles, the transversus abdominis lies under the internal oblique. Its name derives from the direction of its fibers, which pass horizontally across the lateral abdomen to an aponeurosis and the linea alba. The length of the fibers varies considerably depending on the insertion site, the most inferior to the pubis being the longest. At its origin on ribs 7 to 12, the muscle interdigitates with similar diaphragmatic digitations separated by a narrow tendonous raphe. Attempting to draw the abdominal contents posteriorly toward the spine, effectively strengthens this muscle. Clinical considerations are similar to those of the external and internal obliques.

EXTERNAL INTERCOSTALS
(left) Anterior View

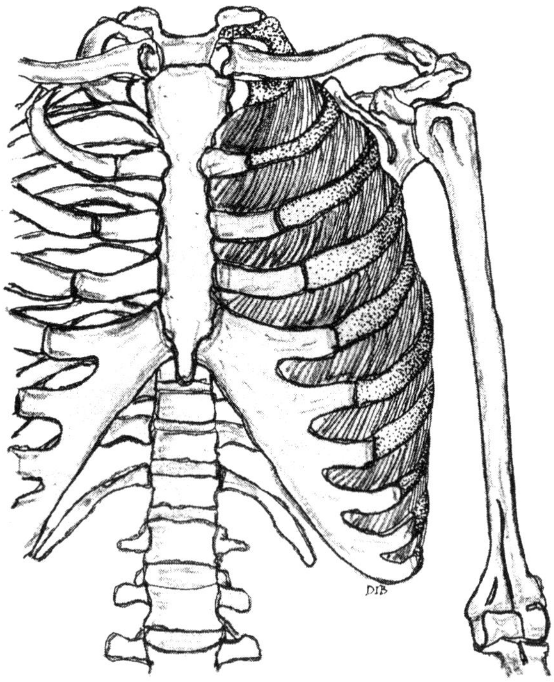

Origin:
Inferior border of rib above

Insertion:
Superior border of rib below

Action:
Elevates ventral portion of rib, increasing the volume of the thoracic cavity

Innervation:
Intercostal nerves (via spinal nerves T2 - T12)

Relationship:
Muscle fibers of the external intercostals are directed obliquely inferiorly and anteriorly. They act with contraction of the diaphragm to aid in inspiration.

INTERNAL INTERCOSTALS
(left) Anterior View

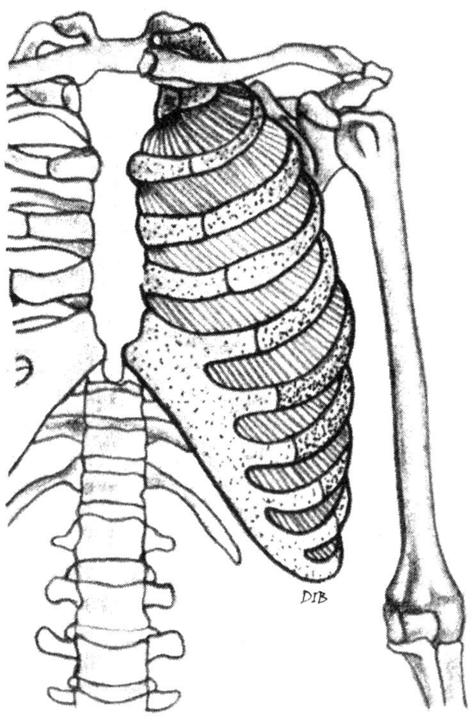

Origin:
Superior border of rib below

Insertion:
Inferior border of rib above

Action:
Depresses ventral portion of rib, decreasing the volume of the thoracic cavity

Innervation:
Intercostal nerves (via spinal nerves T2 - T12)

Relationship:
Muscle fibers of the internal intercostals are directed obliquely inferiorly and posteriorly (at 90° angles to those of the external intercostals). They aid in forced expiration.

TRANSVERSE THORACIS
(left) Anterior View

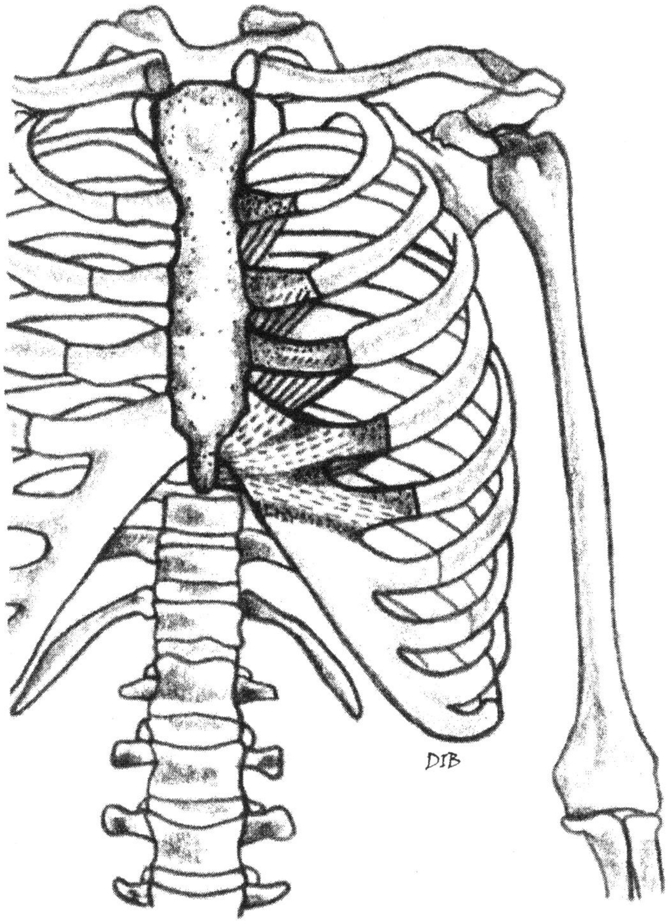

Origin:
Inner surface of the lower aspect of sternum and adjacent costal cartilage

Insertion:
Inner surface of the costal cartilage of ribs 2 thru 6

Action:
Depresses ventral portion of rib, decreasing the volume of the thoracic cavity

Innervation:
Intercostal nerves (via adjacent spinal nerves)

Relationship:
The transverse thoracis are the least developed and the deepest of the costal muscles. The fibers of these muscles vary in their direction, with some being directed obliquely and others horizontally. The muscles are also varied in their presence, occasionally appearing only on one side in the same person. They are believed to act with the internal intercostals and abdominal muscles to assist in forced expiration.
*Some authors also make reference to the innermost intercostal muscles, whose fibers and actions imitate those of the internal intercostals and transverse thoracis.

ANTERIOR ABDOMINAL WALL

(Transverse Section - Superior to Umbilicus)
Superior View

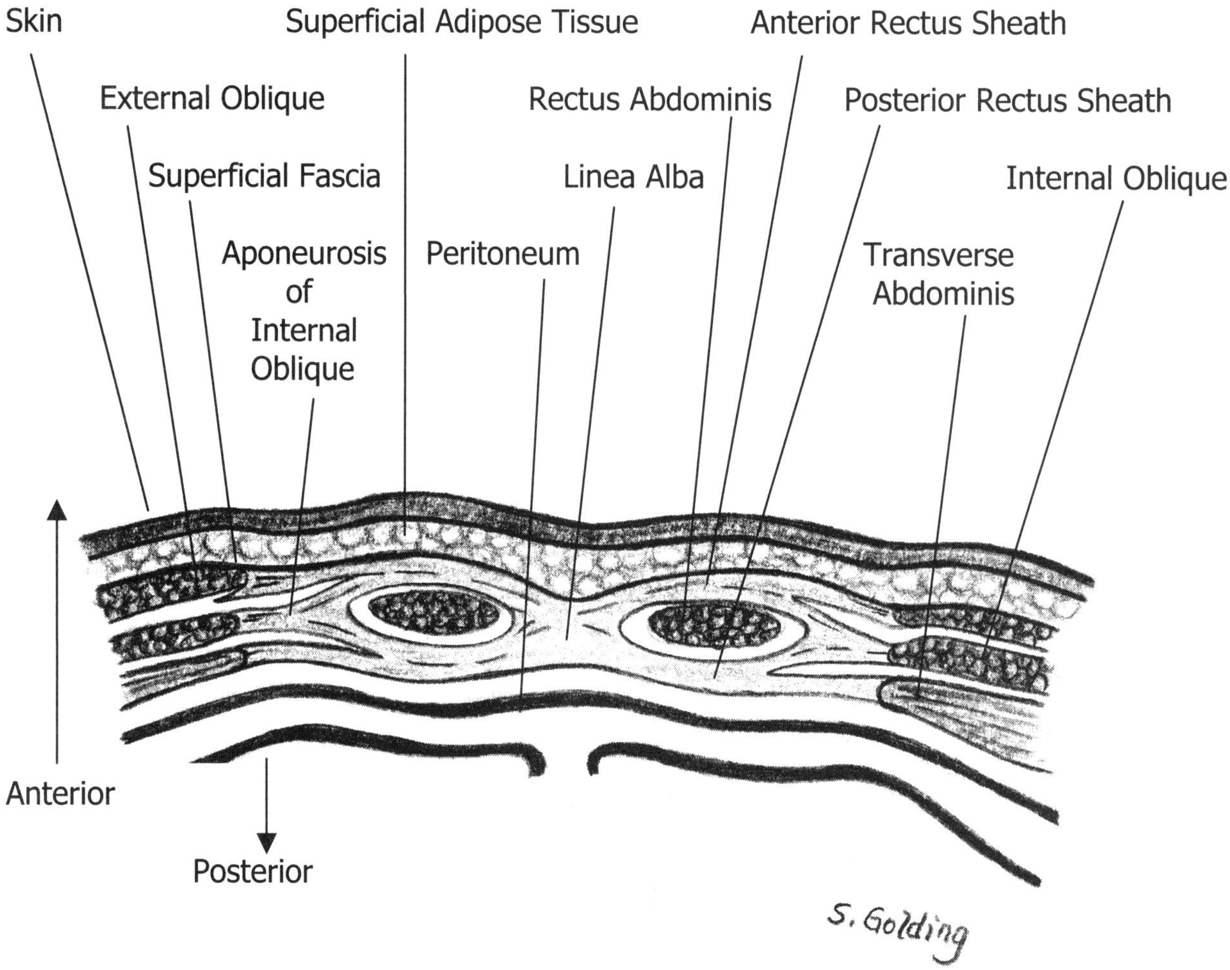

The superior view of a transverse section of the anterior abdominal wall demonstrates the relationship of the numerous associated structures. From superficial to deep (laterally), the structures are: skin, superficial adipose, superficial fascia, fascia and external oblique muscle, fascia and internal oblique muscle, fascia and transverse abdominis muscle, and the peritoneum. From superficial to deep (medially), the structures are: skin, superficial adipose, superficial fascia, fascia (sheath) and rectus abdominis muscle, and the peritoneum. Note that medial to the rectus abdominis muscles lies the linea alba, and lateral to each rectus abdominis muscle, is an area with only skin, superficial adipose, superficial fascia, the aponeurosis of the oblique and transverse muscles, and the peritoneum.

MUSCLES of the PELVIC FLOOR
(Female) Inferior View

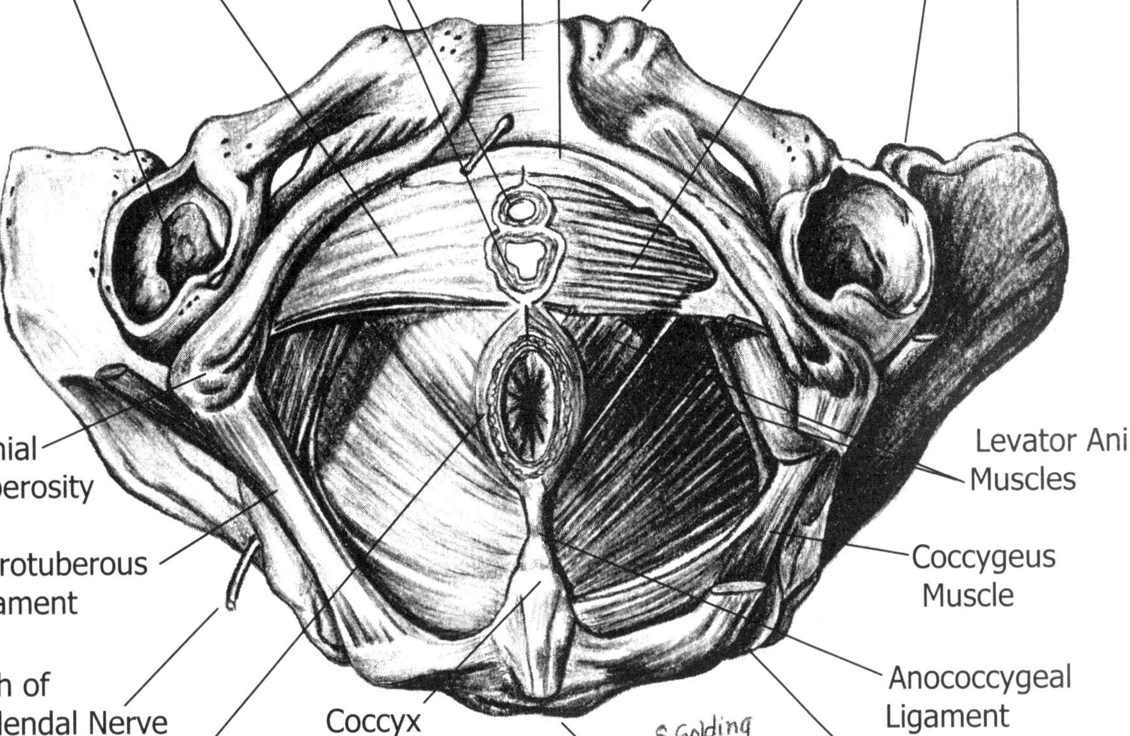

The muscles of the pelvic floor are known collectively as the pelvic diaphragm. This floor is comprised primarily of the levator ani and coccygeus muscles. It stretches from the pubis anteriorly to the coccyx posteriorly and from pelvic wall to pelvic wall laterally. The structure is pierced by the urethra, the anal canal, and by the vagina in females. The pelvic floor muscles function to support the pelvic contents and (with the sphincter muscles) to help provide fecal and urinary continence. Trauma, or childbirth (with prolonged labor) may compromise neuromuscular function and lead to stress incontinence. Kegel exercises with sEMG biofeedback have been shown to very successful in strengthening these muscles.

CORACOBRACHIALIS (left)
Anterior View

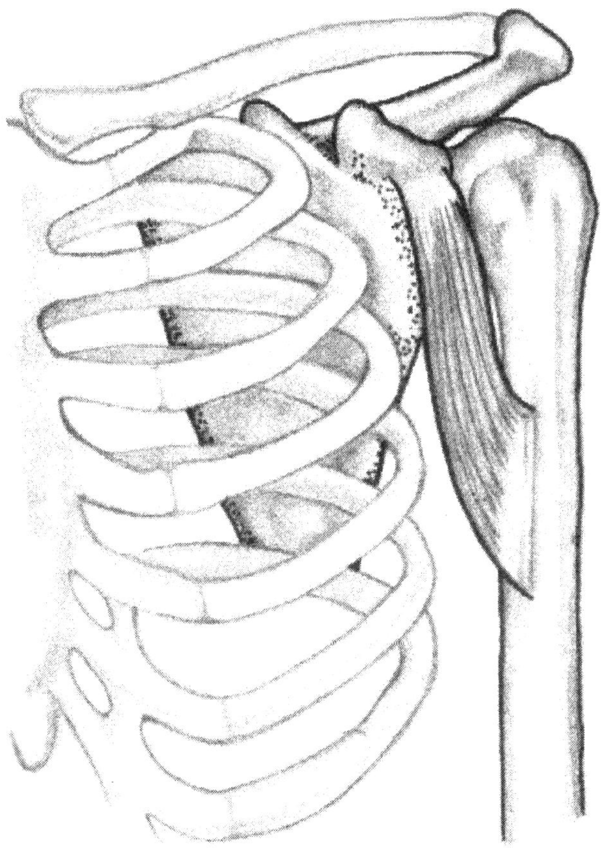

Origin:
Coracoid process (apex) of scapula (from a common tendon of the short head of biceps brachii), Intermuscular septum

Insertion:
Humerus (midway along medial surface of shaft, between the attachments of triceps & brachialis)

Action:
Flexion of arm forward and medially at the glenohumeral joint, Adduction of arm

Innervation:
Musculocutaneous nerve (via spinal nerves C-5, C-6, & C-7)

Relationship:
Coracobrachialis is the smallest of the arm flexors. It lies along the upper medial portion of the humerus and appears as a small rounded ridge. Anterior to coracobrachialis is found pectoralis major. Posterior are found the tendons of subscapularis, latissimus dorsi, and teres major. Coracobrachialis acts synergistically with the anterior deltoid, pectoralis major, and the short head of the biceps in flexion and adduction of the humerus. It is opposed in these actions by the posterior deltoid, latissimus dorsi, teres major, and the long head of the triceps. It is active in swimming and especially in heavy lifting with arms outstretched in front of the torso. The primary symptoms associated with coracobrachialis are localized upper limb pain (felt both anteriorly & posteriorly) and referred pain down the arm to the hand. Pain is aggravated by reaching behind the torso and across the lower back. Trigger point pain is referred down the dorsal aspect of the arm to the hand. Symptoms must be differentiated from cervical neuralgia, carpal tunnel syndrome, tendonitis, subacromial bursitis, and acromioclavicular joint dysfunction.

BICEPS BRACHII (left)

Anterior View

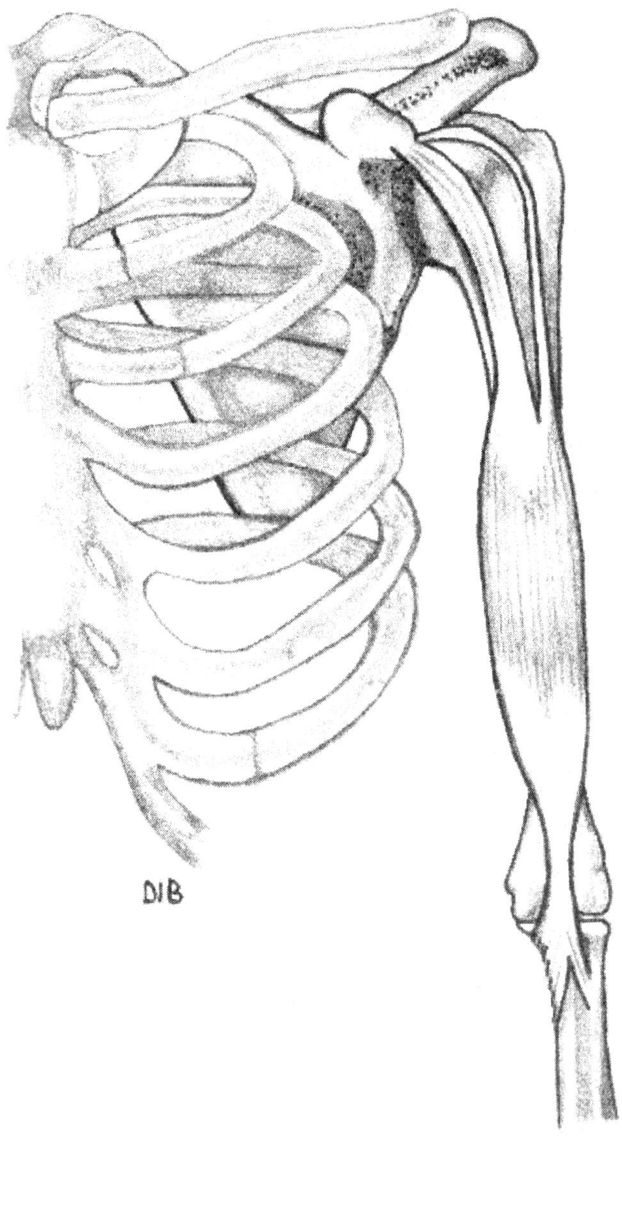

Origin:
Short head: Coracoid process (apex) of scapula
Long head: Supraglenoid tubercle of scapula

Insertion:
Radius (rough surface of radial tuberosity)
Broad bicipital aponeurosis fusing with deep fascia over forearm flexors

Action:
Both heads: Flexion of forearm at the elbow
Supination of forearm (powerful)
Flexion of humerus at the shoulder (weak)
Long head: Stabilizes and depresses humeral head in glenoid fossa during deltoid activity

Innervation:
Musculocutaneous nerve (via spinal nerves C-5 & C-6)

Relationship:
Biceps Brachii is a long muscle on the anterior surface of the arm consisting of two heads and crossing three joints. Its distinctive location helps to explain the complex nature of its actions on the glenohumeral, humeroulnar, and humeroradial joints. Biceps is active in lifting, pull-ups & chin-ups (w/distal attachment fixed), and in the lowering of a weighted object (eccentric contraction). Curls most effectively strengthen it when the hand is supinated. The triceps brachii is its major antagonist. Biceps usually presents with symptoms of local superficial pain, weakness and/or crepitus (upon abduction). Complaints are aggravated by overuse from tennis (backhand), heavy lifting (with the hand supinated), and by repeated forceful supination. Trigger points refer pain to the shoulder and antecubital region. Biceps symptoms must be differentiated from those of arthritis, bursitis, tendonitis, and referred pain of cervical joint dysfunction.

BRACHIALIS (left)
Anterior View

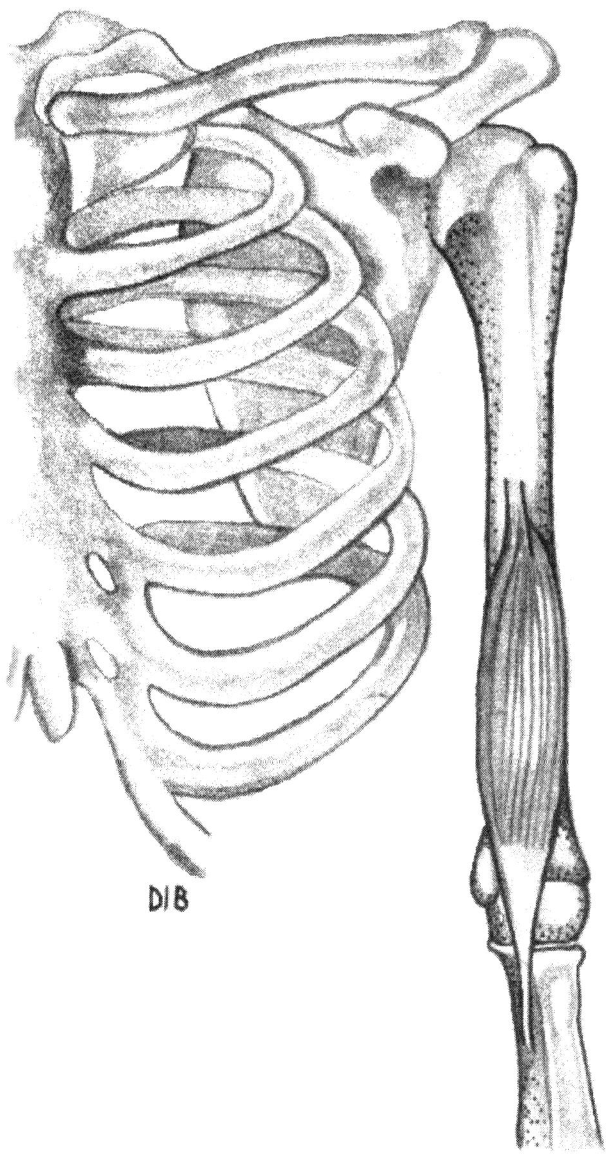

Origin:
Humerus (distal 1/2 of anterior shaft), Intermuscular septum

Insertion:
Ulna (ulnar tuberosity and rough surface of coronoid process), anterior ligament of elbow joint

Action: "True" flexion of ulna at the elbow

Innervation:
Musculocutaneous nerve (via spinal nerves C-5 & C-6)

Relationship:
Brachialis is positioned over the distal half of the humerus and the anterior aspect of the elbow joint. Brachialis is the only pure flexor and is known as the workhorse of the elbow. It is active in lifting, pull-ups & chin-ups (with the ulna fixed), and lowering of a weighted object (eccentric contraction). It is strengthened by these same actions. Brachialis is synergistic with the biceps brachii, brachioradialis, and supinator. It is antagonistic to the triceps brachii. Complaints are typically brought on by either acute or repetitive overuse (as in carrying heavy objects, using a power tool, relentless ironing, or the incessant playing of a guitar or violin). Symptoms commonly include local & referred pain and paresthesia. Trigger point pain is referred to the antecubital region and thumb. These symptoms must be differentiated from those of tendonitis, carpal tunnel syndrome, and the referred pain of cervical joint dysfunction.

BRACHIORADIALIS (right)
Anterior (neutral) View

Origin:
Humerus (anterior aspect of proximal 2/3 of lateral supracondylar ridge and lateral intermuscular septum

Insertion:
Radius (lateral side of base just proximal to the styloid process)

Action:
Flexion of elbow - Note: This muscle evolved with the extensor muscles and is innervated by the radial nerve, but its action is that of a forearm flexor. The muscle is less active when the forearm is fully supinated because it crosses the joint laterally rather than anteriorly

Innervation:
Radial nerve (via spinal nerves C-5 & C-6, with C7 innervation sometimes cited)

Relationship:
Brachioradialis is the most superficial muscle on the radial side of the forearm and forms the lateral side of the cubital fossa. It has a rather thin belly, which descends to the midforearm, where its long flat tendon begins and continues to the distal radius. Brachioradialis acts synergistically with biceps brachii and brachialis, although, it excels during speedy movements of the elbow. It is opposed by triceps brachii. Curls, with the radioulnar joint in neutral position, are best suited to strengthen brachioradialis. Symptoms associated with brachioradialis include lateral epicondyle pain, weakness, and writer's cramp, all of which are aggravated by repeated and extended use. Trigger points refer pain to the lateral epicondyle and downward to the wrist and thumb. Symptoms must be differentiated from those of carpal tunnel syndrome, lateral epicondylitis (tennis elbow), arthritis, C-5/C-6 radiculopathy, and carpal or radioulnar joint dysfunction.

SUPINATOR (left)
Dorsal View

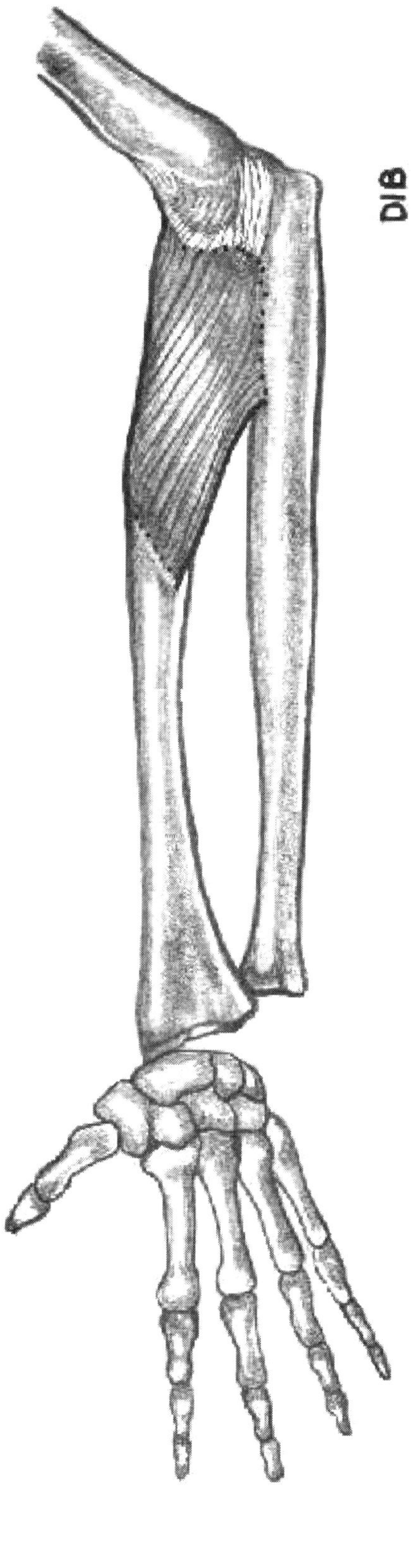

Origin:
Humerus: lateral epicondyle, radial collateral ligament of elbow joint
Ulna (dorsal surface of shaft, supinator crest), annular ligament of radioulnar joint

Insertion:
Radius:
Superficial plane: lateral edge of radial tuberosity
Deep plane: medial and posterior surfaces of the radial tuberosity and dorsal and medial surfaces of upper 1/3 of radial shaft

Action:
Supination of forearm at the radioulnar joint.
(possibly flexion of elbow by fibers at the capsule)

Innervation:
Radial nerve (via spinal nerves C-5 & C-6)

Relationship:
Supinator is a broad muscle whose fibers form two planes that curve around the upper radius. The two planes arise together from the epicondyle; the superficial plane from the tendon and the deep plane from the muscle fibers. This muscle is subject to considerable variation. It is the prime supinator of the extended forearm, but is synergistic with biceps brachii in supination when the elbow is even slightly flexed. Its chief antagonists are pronator quadratus and teres. Supinator is active in the twisting and screwing type movements of tennis, opening doors & jars, and carrying heavy objects with the elbow extended. It is best strengthened by supinating against resistance with the elbow in full extension. Symptoms include pain in the lateral epicondyle and dorsal surface of hand. Complaints are aggravated by overuse during the above-mentioned activities. Supinator is commonly involved in tennis elbow. Trigger points refer pain to the lateral epicondyle and down to the thumb. Symptoms must be differentiated from those of lateral epicondylitis, arthritis, and cervical joint dysfunction.

PRONATOR TERES (left)
Palmer view

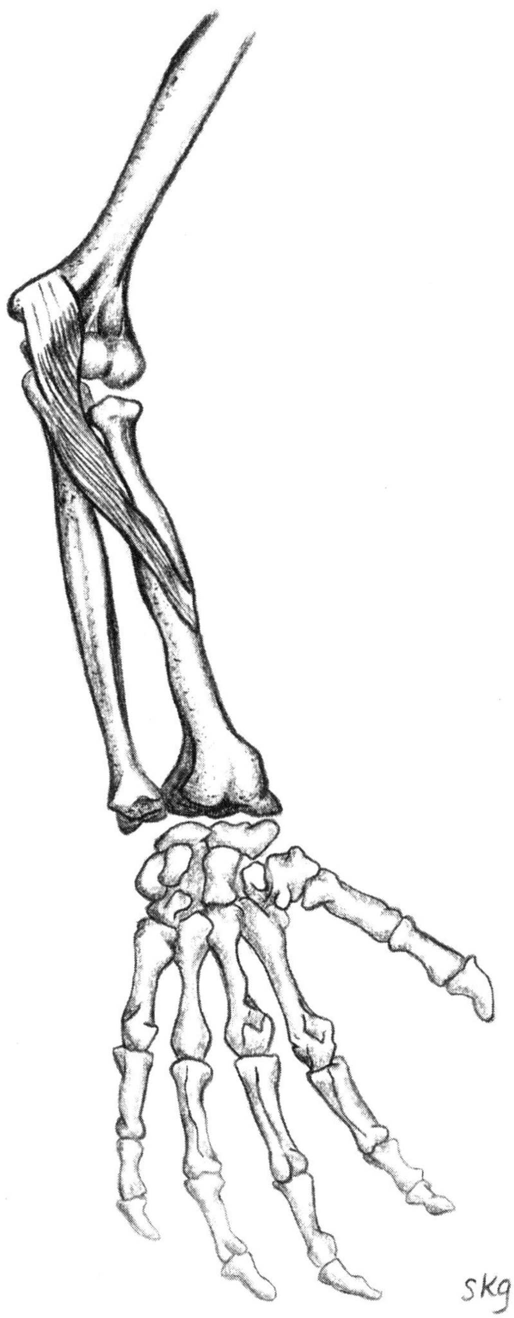

Origin:
Humeral head (superficial head): Distal supracondylar ridge, medial epicondyle, and common tendon of origin of flexor muscles
Ulnar head (deep head): Coronoid of medial ulna

Insertion:
Radius (lateral aspect, middle portion of shaft)

Action:
Pronation of forearm at the radioulnar joint,
Elbow flexion (assists, when resisted)

Innervation:
Median nerve (via spinal nerves C-6 & C-7)

Relationship:
In pronator teres, the larger head is the humeral. The smaller, thin ulnar head joins its companion at an acute angle. Together they pass obliquely across the forearm to end in a flat tendon of insertion near the radius. The lateral border of this muscle is the medial limit of the cubital fossa, which lies just anterior to the elbow joint. It is synergistic with pronator quadratus, which is the primary pronator of the forearm at the radioulnar joint. It is opposed in pronation by supinator. It is best strengthened by pronating against resistance with the elbow in full extension. Symptoms associated with pronator teres include local and referred pain, weakness, and the inability to supinate the cupped hand. Complaints are aggravated by repeated, intense gripping, movements such as grasping a tool, a steering wheel, or ski poles for extended periods of time. Trigger points refer pain to the radial aspect of forearm down to the wrist. Symptoms must be differentiated from those of medial nerve neuropathy, medial epicondylitis, arthritis, carpal tunnel syndrome, and cervical joint dysfunction.

PRONATOR QUADRATUS (left)
Palmer View

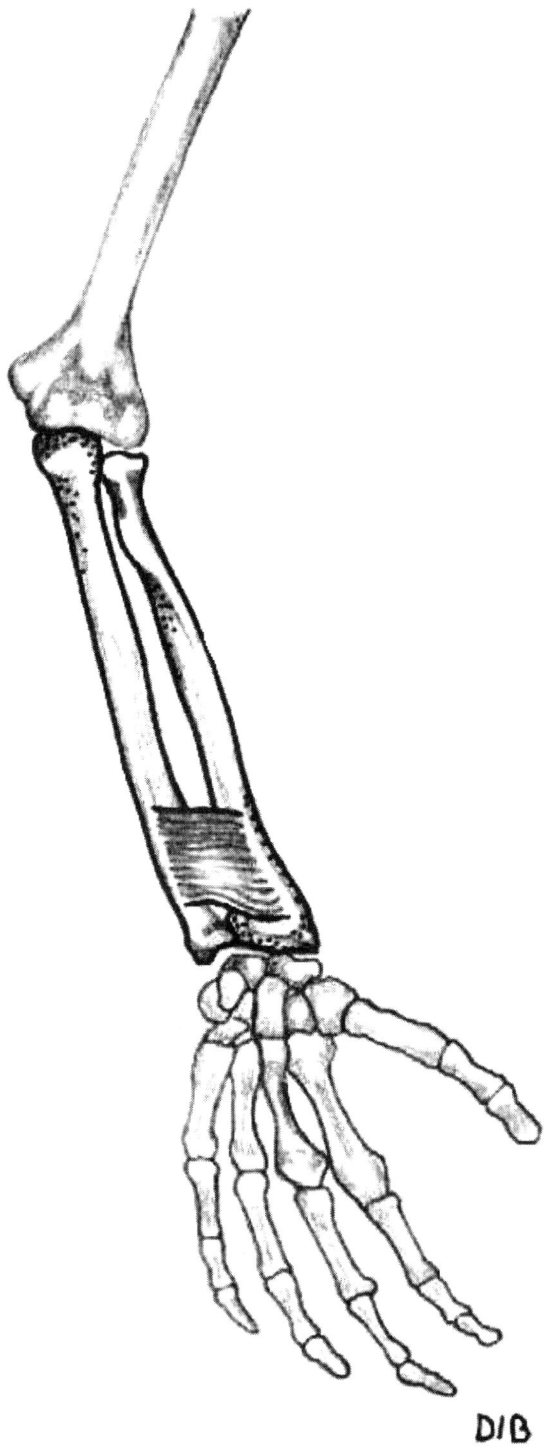

Origin:
Ulna (anterior and medial surfaces of distal 1/4 and oblique ulnar ridge on body),
Aponeurosis over middle 1/3 of the muscle

Insertion:
Radius (anterior surface of distal 1/4 of shaft; deeper
fibers to a narrow triangular area above ulnar notch)

Action:
Pronation of forearm at the radioulnar joint

Innervation:
Median nerve (via spinal nerves C-8 & T-1)

Relationship:
Pronator quadratus is a small, flat, quadrilateral muscle that passes across the anterior aspect of the distal ulna to the distal radius. Its fibers are quite horizontal. Pronator quadratus is the deepest muscle of the distal forearm. It is the primary pronator of the forearm at the radioulnar joint.
It is synergistic with pronator teres (during rapid or forceful pronation) and is opposed by supinator.
It is best strengthened by pronating against resistance with the elbow in full extension.
Symptoms associated with pronator quadratus include local and referred pain and weakness.
Complaints are aggravated by repeated, intense gripping and twisting movements. Trigger point pain is referred to the palmar surface of the wrist.
Symptoms must be differentiated from those of medial nerve neuropathy, medial epicondylitis, arthritis, carpal tunnel syndrome, and cervical joint dysfunction.

PALMARIS LONGUS (left)
Palmer View

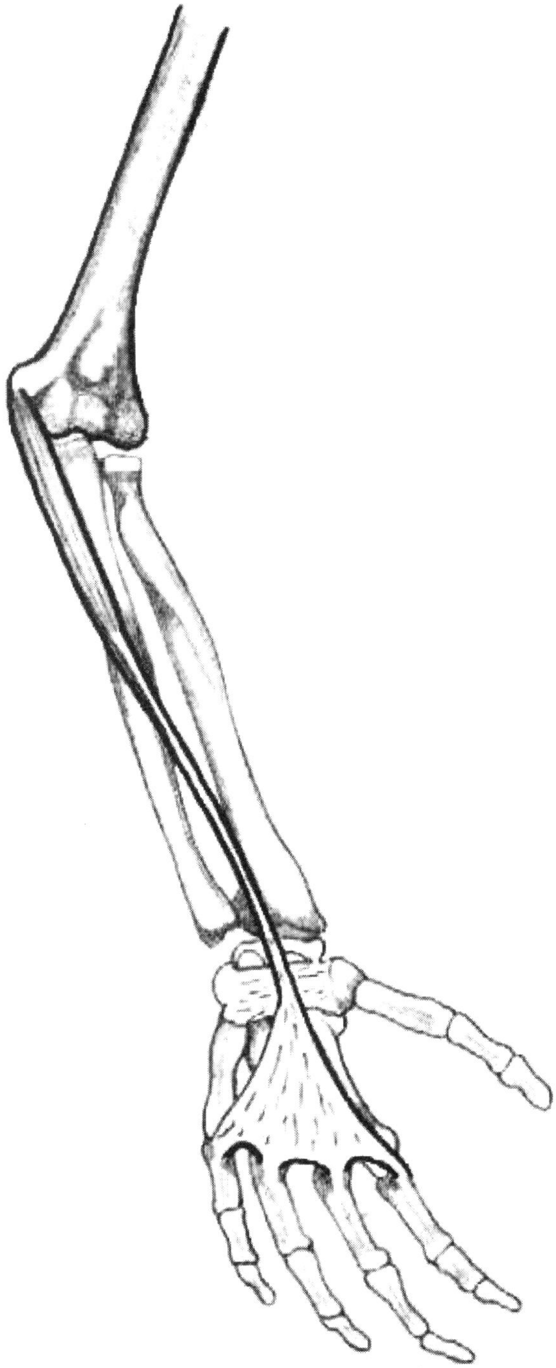

Origin:
Humerus (medial epicondyle via common flexor tendon), Intermuscular septum

Insertion:
Flexor retinaculum (distal), Palmar aponeurosis, with a slip sent frequently to the short thumb muscles

Action:
Tension of palmar fascia (cupping the palm)
Flexion of hand at wrist (weak)
Flexion of elbow at the humeroulnar joint (weak)

Innervation:
Median nerve (via spinal nerves C-6 & C-7)

Relationship:
Palmaris longus is a slim fusiform muscle. Its belly Is located in the proximal forearm and ends in a long tendon midway in the forearm. The tendon traverses the carpal retinaculum within a synovial sheath. This muscle is quite variable and may be absent (bilaterally) in up to 20% of the population. It functions with flexor carpi ulnaris and flexor digitorum superficialis in gripping. Palmaris longus may be strengthened by wrist curls against resistance. Symptoms most often include needle-like prickling pain in the palm that is made worse by repetitive activity, and contracture of the palmar fascia. Pain may also radiate proximately. Trigger points refer pain to the wrist and palm. Symptoms must be differentiated from those of true carpal tunnel syndrome, tennis elbow, and arthritis.

FLEXOR CARPI RADIALIS (left)
Palmer View

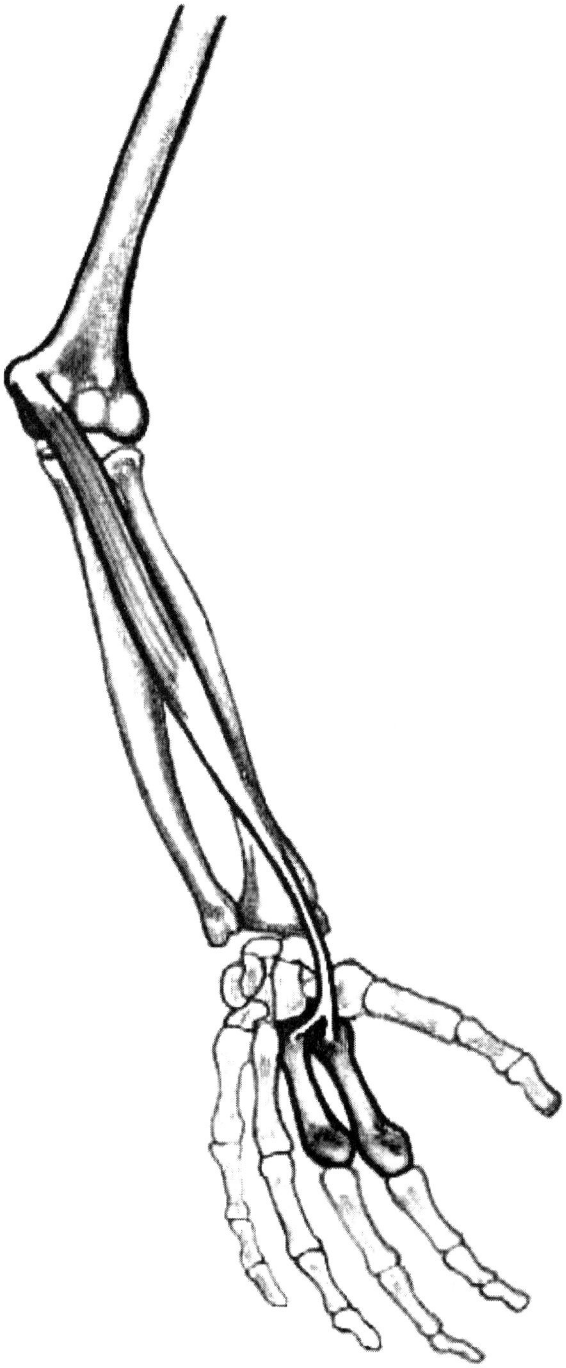

Origin:
Humerus (medial epicondyle by common flexor tendon), intermuscular septa, antebrachial fascia

Insertion:
2nd and 3rd metacarpals (bases)

Action:
Flexion of wrist radial deviation of wrist,
Extend fingers,
Flexion of elbow at the humeroulnar joint (weakly assists),
Pronation of forearm (weakly assists)

Innervation:
Median nerve (via spinal nerves C-6 & C-7)

Relationship:
Flexor carpi radialis is a slender, aponeurotic muscle at its origin. It descends in the forearm between the pronator teres and the palmaris longus. It increases in size as it descends to end in a tendon about halfway down the forearm. Flexor carpi radialis may be strengthened by resisted wrist curls. Symptoms often include pain and tenderness at the volar aspect of the wrist that is made worse by repetitive and forceful gripping. Pain near the medial epicondyle has also been reported. The painless locking of the fingers in flexed position (trigger finger) may be related to active trigger points in this muscle. Trigger points refer pain to the wrist and palm. These symptoms should be differentiated from those of medial epicondylitis, carpal tunnel syndrome, ulnar neuropathy, arthritis, and cervical radiculopathy.

FLEXOR DIGITORUM SUPERFICIALIS (left)
Palmer View

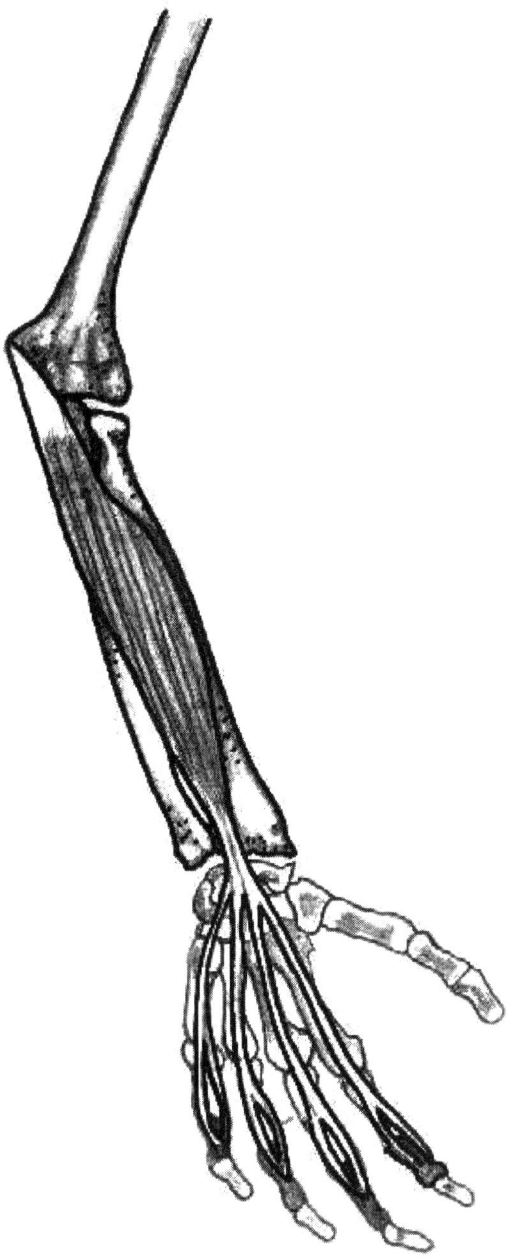

Origin:
Humeral-ulnar head: Humerus (medial epicondyle via the common flexor tendon), ulna (ulnar collateral ligament of elbow joint) and intermuscular septum
Radial head: Radius (oblique line on anterior surface of shaft)

Insertion:
Middle phalanges (digits 2 to 5), via 4 tendons

Action:
Flexion of PIP joints of digits 2 to 5, Flexion of MP joints of digits 2 to 5 (assists)
Flexion of wrist (assists, especially in forceful grasp)

Innervation:
Median nerve (via spinal nerves C-7 & C-8)

Relationship:
Flexor digitorum superficialis lies deep to the other forearm flexors but is the largest superficial flexor. The muscle separates into two planes of fibers: superficial and deep. The superficial plane (joined by the radial head) divides into two tendons for the middle and ring fingers. The deep plane fibers divide and join the tendons to the index and little fingers. The four tendons sweep under the flexor retinaculum arranged in pairs for the middle and ring fingers, and for the index and little fingers. The tendons diverge again in the palm. At the base of the proximal phalanges, each divides into two slips, permitting passage of the flexor digitorum profundus to each finger. The slips reunite and then divide again for a final time to insert on both sides of each middle phalanx. The radial head may be absent. It is strengthened by gripping, squeezing, and clenching types of activities. Symptoms often include pain and tenderness at the volar aspect of the wrist that is made worse by repetitive and forceful gripping. Trigger points refer pain to the palm and fingers. These symptoms must be differentiated from those of medial epicondylitis, carpal tunnel syndrome, ulnar neuropathy, osteoarthritis, and cervical radiculopathy.

FLEXOR DIGITORUM PROFUNDUS (left)
Palmer View

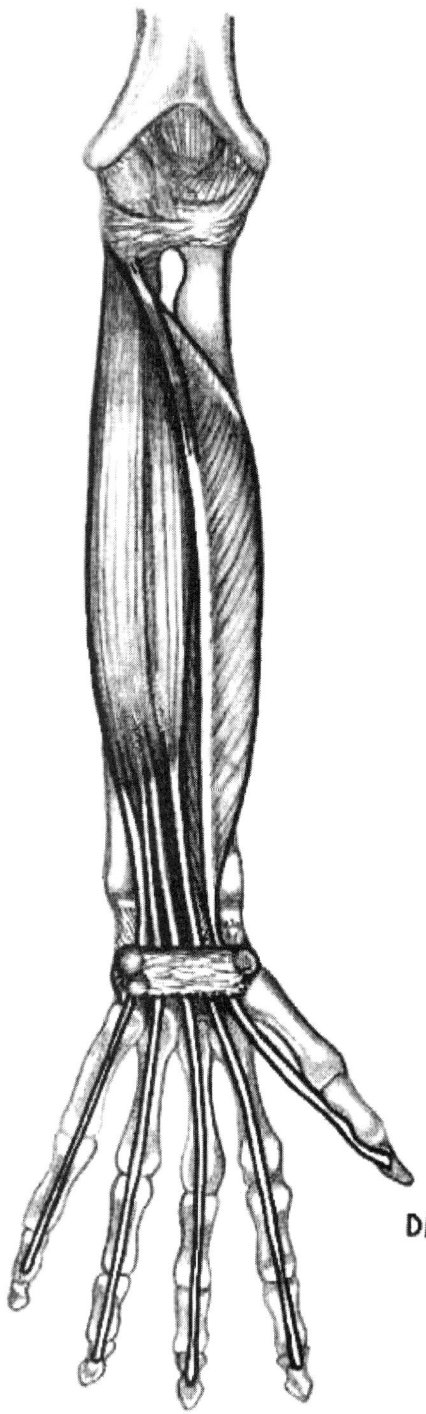

Origin:
Ulna (upper 3/4 of anterior and medial surfaces of body and also coronoid process),
Interosseous membrane

Insertion:
Splits into four tendons for digits 2 to 5, distal phalanges and base of digits 2 to 5. Each tendon passes through openings in the continuous tendon of the flexor digitorum superficialis and inserts on the base of the distal phalanx of each finger

Action:
Flexion of DIP joints of digits 2 to 5
Flexion of MP and PIP joints of digits 2 to 5 (assists)
Flexion of wrist (assists)

Innervation:
Median nerve (via C8-T1, for digits 2 and 3)
Ulnar nerve (via C8-T1, for digits 4 and 5)

Relationship:
Profundus is situated on the ulnar side of the forearm, deep to the superficial flexors. The muscle fibers end in four tendons below the midforearm. These tendons pass into the hand under the transverse carpal ligament. The tendon for the index finger remains distinct, but the tendons for the other fingers are interwoven with areolar tissue and tendinous slips well down into the palm. After passing through the tendons of the flexor digitorum superficialis, they move to their insertions. The four lumbrical muscles arise with the profundus in the palm. Flexor digitorum profundus is strengthened by gripping, squeezing, and clenching types of activities. Symptoms often include pain and tenderness at the volar aspect of the wrist, which is made worse by repetitive and forceful gripping. Trigger points refer pain to the palm and fingers. These symptoms must be differentiated from those of medial epicondylitis, carpal tunnel syndrome, ulnar neuropathy, osteoarthritis, and cervical radiculopathy.

FLEXOR POLLICIS LONGUS (left)
Palmer View

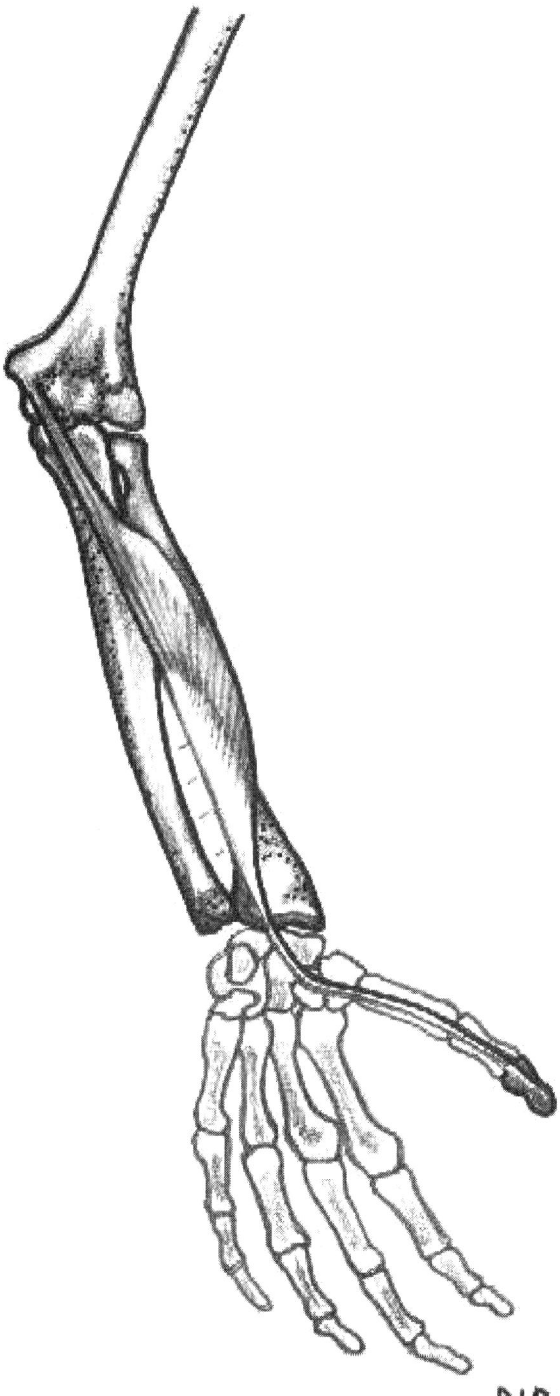

Origin:
Radius (grooved anterior surface of middle half of body),
Interosseous membrane,
Ulna (coronoid process) - variable,
Humerus (medial epicondyle) - variable

Insertion:
Thumb (base of distal phalanx, palmar surface)

Action:
Flexion of IP joint of thumb,
Flexion of the MP and CMC joints of thumb (assists),
Flexion of wrist (assists)

Innervation:
Median nerve (anterior interosseous branch, via spinal nerves C-8- & T-1)

Relationship:
Flexor pollicis longus descends on the radial side of the forearm in the same plane, but lateral, to the flexor digitorum profundus. Flexor pollicis longus may be strengthened by gripping, squeezing, and clenching types of activities that emphasize thumb opposition. Symptoms associated with flexor pollicis longus often include pain and tenderness at the volar aspect of the wrist that is made worse by repetitive and forceful gripping. Trigger points refer pain to the wrist and thumb. These symptoms must be differentiated from those of medial epicondylitis, carpal tunnel syndrome, ulnar neuropathy, osteoarthritis, and cervical radiculopathy.

FLEXOR CARPI ULNARIS (left)
Palmer View

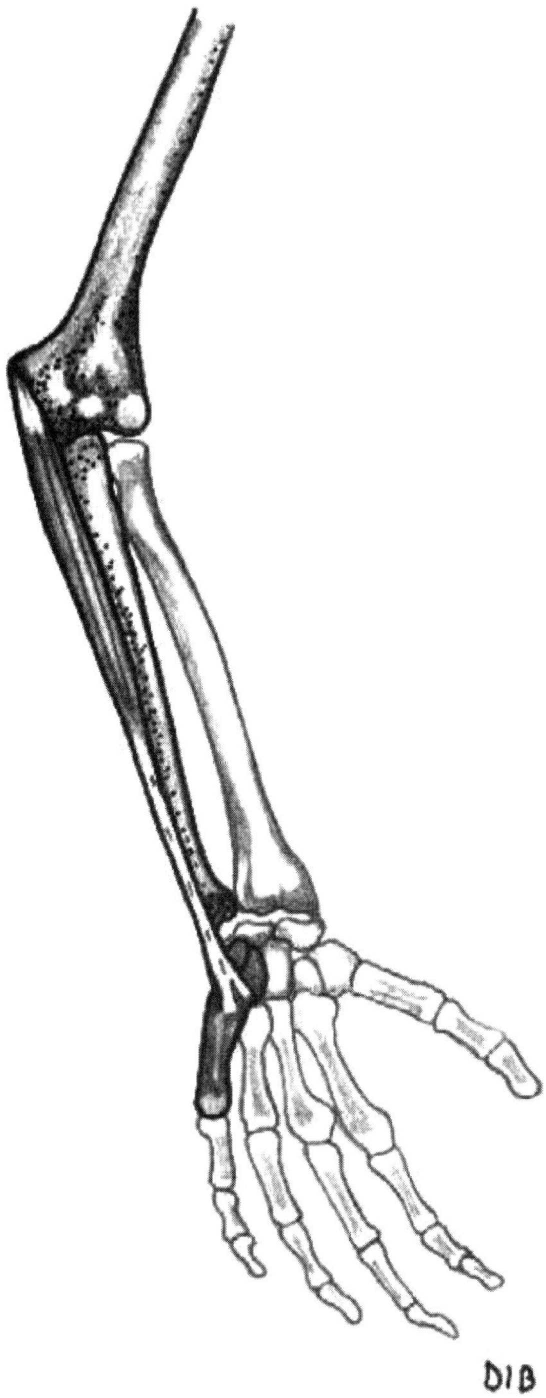

Origin:
Humeral head: Humerus (medial epicondyle via common flexor tendon).
Ulnar head: Ulna (olecranon, medial process, and upper 2/3 of body via an aponeurosis), Intermuscular septum

Insertion:
Pisiform bone, Hamate bone, 5th metacarpal

Action:
Flexion of wrist, ulnar deviation of wrist, flexion of elbow at the humeroulnar joint (assists)

Innervation:
Ulnar nerve (via spinal nerves C-8 & T-1)

Relationship:
Flexor carpi ulnaris is the most ulnar-lying of the flexors in the forearm. The humeral head is small in contrast to the extensive origin of the ulnar head. The two heads are connected by a tendinous arch under which the ulnar nerve descends. The muscle fibers end in a tendon that forms along the anterolateral border of the muscle's distal half. Flexor carpi ulnaris may be strengthened by wrist curling exercises with resistance. Symptoms associated with flexor carpi ulnaris often include pain and tenderness at the volar aspect of the wrist that is made worse by repetitive and forceful gripping. Trigger points refer pain to the ulnar aspect of the wrist. These symptoms must be differentiated from those of medial epicondylitis, carpal tunnel syndrome, ulnar neuropathy, osteoarthritis, and cervical radiculopathy.

EXTENSOR CARPI RADIALIS LONGUS (left)
Dorsal View

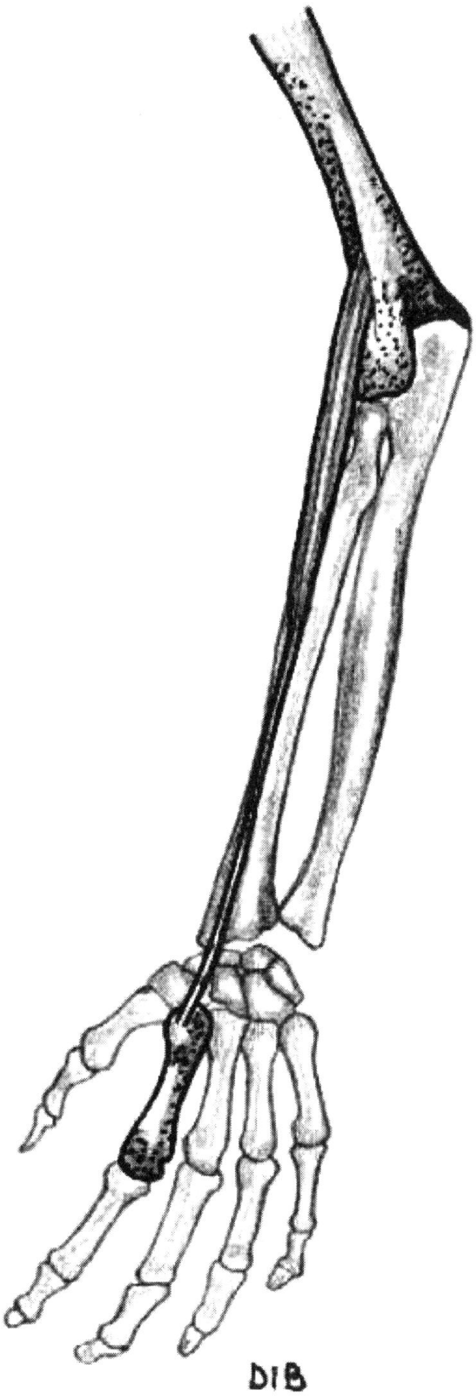

Origin:
Humerus (distal 1/3 of lateral supracondylar ridge),
Lateral intermuscular septum,
Common extensor tendon

Insertion:
2nd metacarpal (dorsal surface of base on radial side),
Occasionally slips to 1st and 3rd metacarpals

Action:
Extension and radial deviation of wrist,
Synergist for finger flexion by stabilization of wrist, assists in elbow flexion

Innervation:
Radial nerve (via spinal nerve C-6, with C5 and C7 also being reported)

Relationship:
Extensor carpi radialis longus descends lateral to brachioradialis. Its muscle fibers end at the midforearm in a flat tendon. Performing wrist extensions against resistance strengthens this muscle. Symptoms associated with extensor dysfunction most often include pain noted at the wrist, weakness, and decreased ROM. Trigger points refer pain to the dorsal aspect of the forearm and wrist. These symptoms should be differentiated from those of tennis elbow, arthritis, and problems originating with serratus posterior and the scalenes.

EXTENSOR CARPI RADIALIS BREVIS (left)

Dorsal View

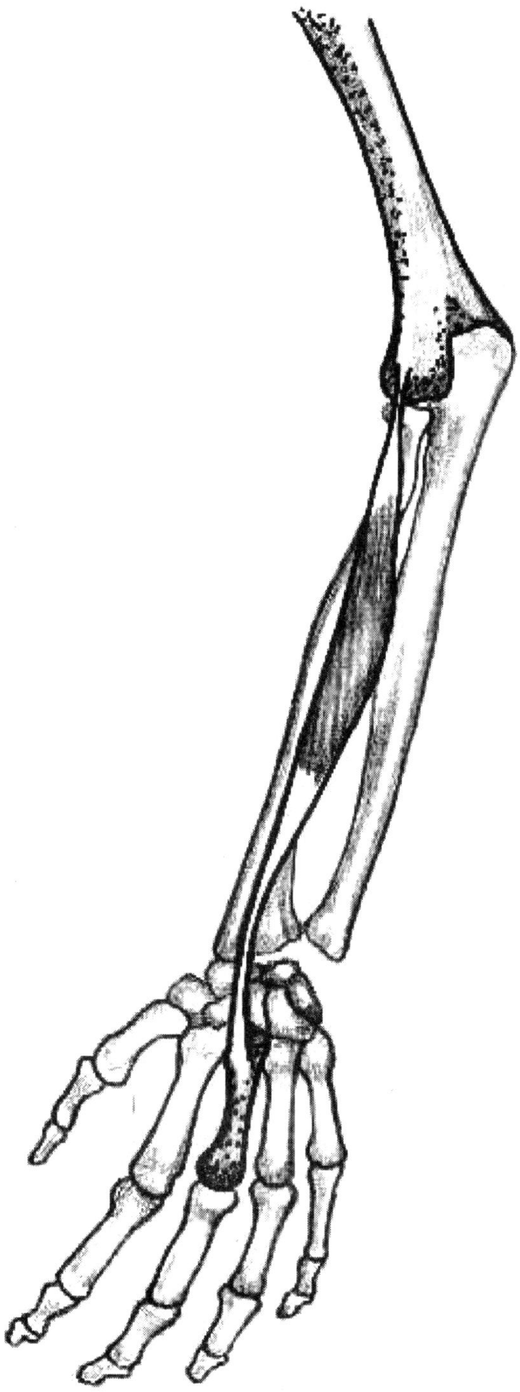

Origin:
Humerus (lateral epicondyle via common extensor tendon),
Radial collateral ligament of elbow joint,
Aponeurosis sheath and intermuscular septum

Insertion:
3rd metacarpal (dorsal surface of base on radial side, on or just distal to styloid process),
Slips often sent to 2nd metacarpal

Action:
Extension of wrist, Radial deviation of wrist (weak),
Finger flexion synergist (by stabilizing the wrist)

Innervation: Radial nerve (via spinal nerves C-6 & C-7)

Relationship:
Extensor carpi radialis brevis is a short, thick muscle that lies partially under the extensor carpi radialis longus in the upper forearm. Its muscle fibers end well above the wrist in a flattened tendon, which descends along the extensor carpi radialis longus tendon to the wrist. Performing wrist extensions against resistance strengthens this muscle. Symptoms associated with extensor dysfunction most often include pain noted at the wrist, weakness, and decreased ROM. Trigger points refer pain to the dorsal aspect of the wrist. These symptoms should be differentiated from those of tennis elbow, arthritis, and problems originating with serratus posterior and the scalenes.

EXTENSOR CARPI ULNARIS (left)
Dorsal View

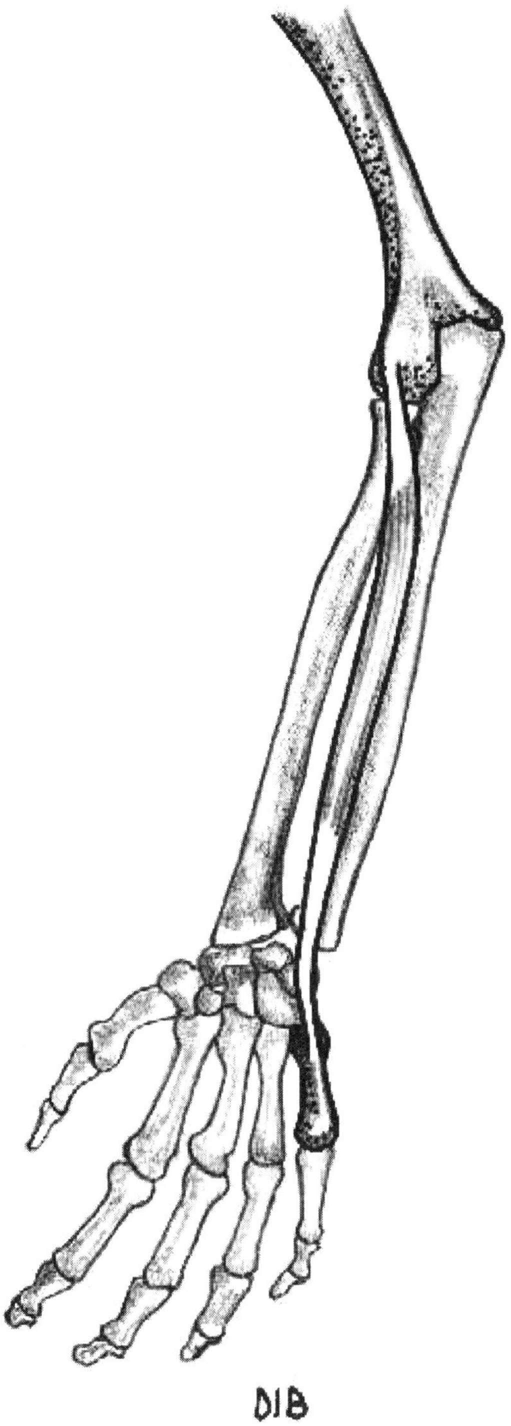

Origin:
Humerus (lateral epicondyle via common extensor tendon),
Ulna (dorsal border with an aponeurosis common to flexor carpi ulnaris and flexor digitorum profundus)

Insertion:
5th metacarpal (tubercle on ulnar side of base)

Action:
Extension of wrist,
Ulnar deviation of hand

Innervation:
Radial nerve (deep branch), (via spinal nerves C-6 & C-8)

Relationship:
Extensor carpi ulnaris muscle fibers descend on the dorsal-ulnar side of the forearm and join a tendon located in the distal 1/3 of the forearm that is the most medial tendon on the dorsum of the hand. This tendon can be palpated lateral to the groove found just over the ulna's posterior border. Performing wrist extensions against resistance strengthens this muscle. Symptoms associated with extensor dysfunction most often include pain noted at the wrist, weakness, and decreased ROM. Trigger points refer pain to the ulnar aspect of the wrist. These symptoms should be differentiated from those of tennis elbow, arthritis, and problems originating with serratus posterior and the scalenes.

EXTENSOR DIGITORUM (left)

Dorsal View

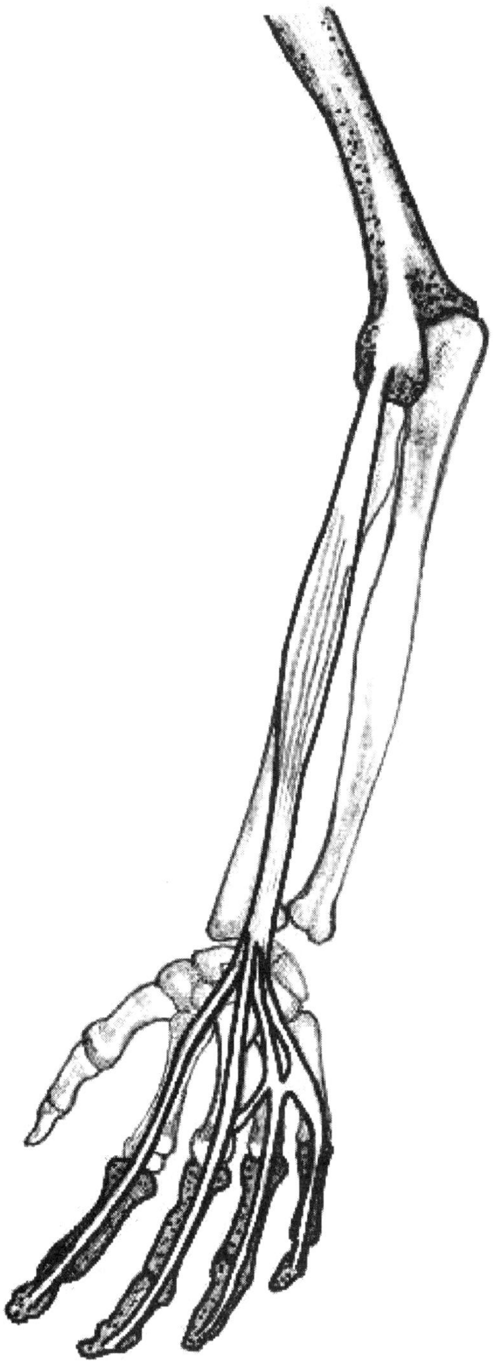

Origin:

Humerus (lateral epicondyle via common extensor tendon), intermuscular septa, antebrachial fascia

Insertion:

Divides distally into four tendons that insert in three variable intertendinous connections on digits 2 thru 5, Intermediate slip: Dorsum of base of each middle phalanx Lateral slips (two): Dorsum of base of each distal phalanx via dorsal expansion

Action:

Extension of MP and IP joints, digits 2 to 5 MP joints (proximal phalanges) directly, IP joints (middle and distal phalanges) indirectly with MP joints flexed wrist extension (assists),

Abduction of ring, index, and little fingers with extension but no such action on the middle finger

Innervation:

Radial nerve (via spinal nerves C-6 thru C-8)

Relationship:

Extensor digitorum divides above the wrist into four distinct tendons, which traverse (with the extensor indices) a tunnel under the extensor retinaculum in a common sheath. Over the dorsum of the hand, the four tendons diverge, one to each finger. The tendon to the index finger is accompanied by the extensor indices tendon. The digital attachments are achieved by a fibrous expansion dorsal to the proximal phalanges. All of the digital extensors, as well as the lumbricales and interossei, are integral to this mechanism. Performing finger extensions against resistance strengthens this muscle. Symptoms associated with extensor dysfunction most often include pain noted at the wrist, weakness, and decreased ROM. Trigger points refer pain to the dorsal aspect of the forearm and hand. These symptoms should be differentiated from those of tennis elbow, arthritis, and problems originating with serratus posterior and the scalenes.

EXTENSOR DIGITI MINIMI (left)
Dorsal View

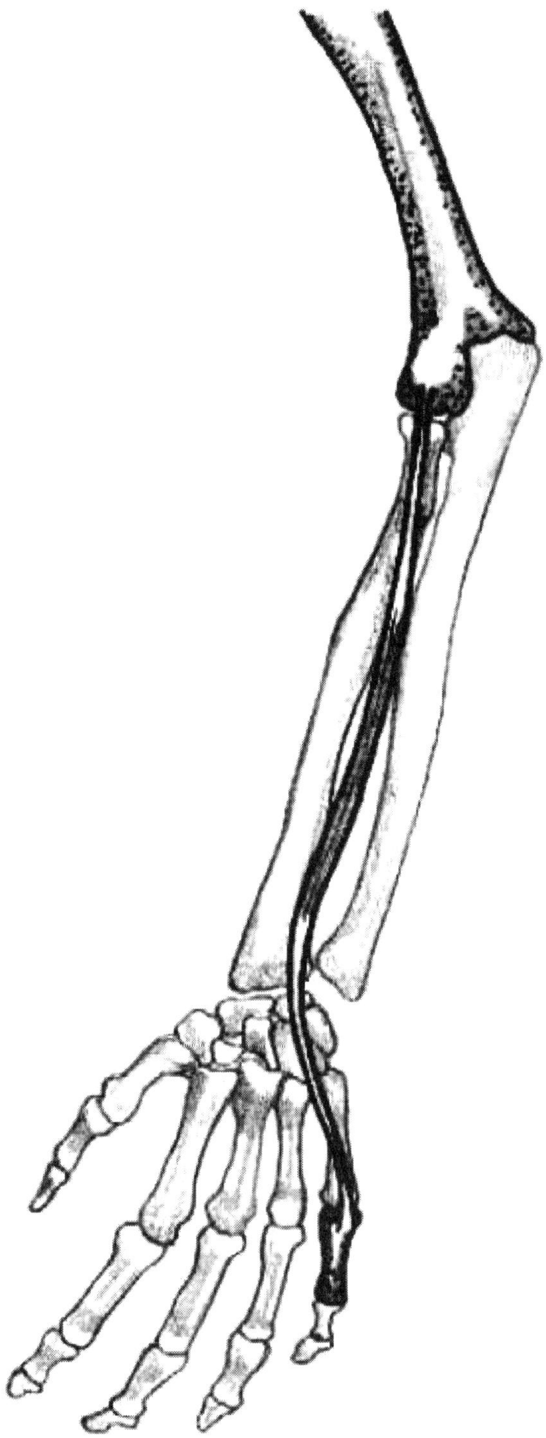

Origin:
Humerus (lateral epicondyle via common extensor tendon),
Intermuscular septum

Insertion:
Proximal phalanx of digit 5 (little finger) on the radial side (proximally divided into two tendons to join the extensor expansion hood), and the tendon of the extensor digitorum

Action:
Extension of MP and IP joints of digit 5 (little finger),
Wrist extension (accessory),
Abduction of digit 5 (assists)

Innervation:
Radial nerve (deep branch), (via spinal nerve roots C-6 thru C-8)

Relationship:
Extensor digiti minimi is a slim, extensor muscle that lies medial to the extensor digitorum. It descends in the forearm between the extensor digitorum and the extensor carpi ulnaris, passes under the extensor retinaculum in its own compartment, and then divides into two tendons. The lateral tendon joins directly with the tendon of the extensor digitorum. All three join the extensor expansion and insert on the proximal phalanx of digit 5. Performing finger extensions against resistance strengthens this muscle. Symptoms associated with extensor dysfunction most often include pain noted at the wrist, weakness, and decreased ROM. Trigger points refer pain to dorsum of the hand and little finger. These symptoms should be differentiated from those of tennis elbow, arthritis, and problems originating with serratus posterior and the scalenes.

EXTENSOR INDICIS (left)
Dorsal View

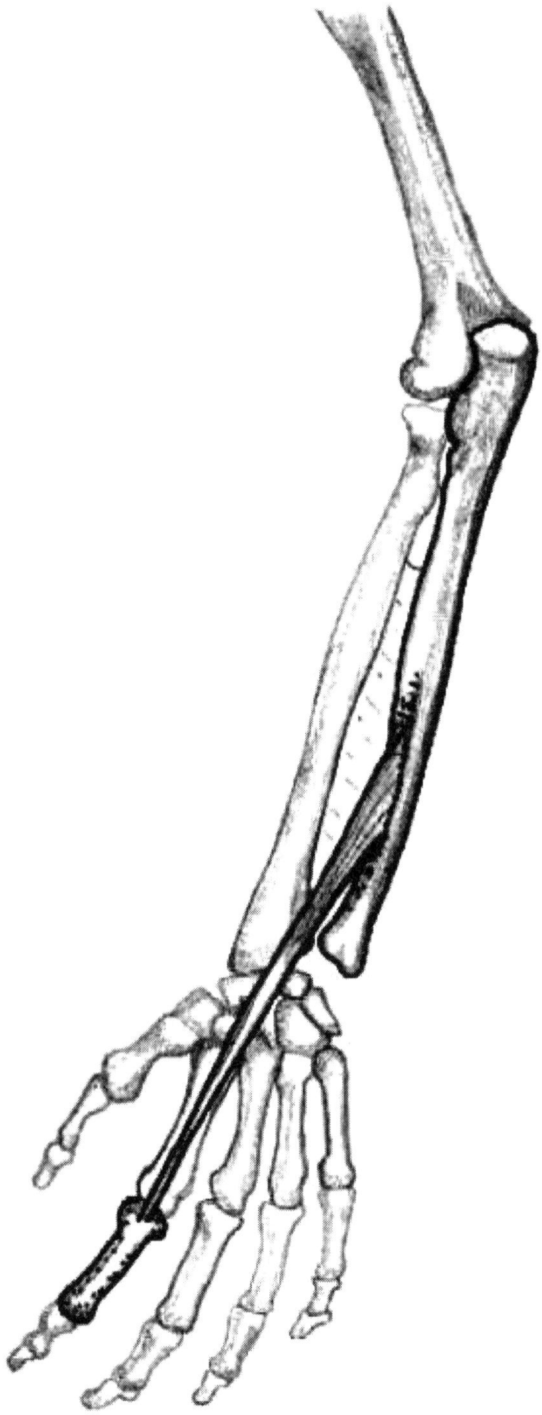

Origin:
Dorsal and lateral surfaces of the ulna below the origin of pollicis longus membrane, Interosseous membrane

Insertion:
Ulnar aspect of the proximal portion of the dorsal surface (proximal phalanx) of the 2nd digit

Action:
Extension of MP joint of index finger,
Extension of IP joints (with intrinsics),
Adduction of index finger (assists),
Wrist extension (assists)

Innervation:
Radial nerve (via spinal nerve roots C6 thru C8)

Relationship:
Extensor indicies arises just below the extensor pollicis longus and travels adjacent with it down to the level of the wrist. It passes under the extensor retinaculum, in a compartment with the tendons of the extensor digitorum. Performing finger extensions against resistance strengthens this muscle. Symptoms associated with extensor indicies dysfunction seldom occur alone and include weakness in extension, limited ROM in flexion, fatigue with repetitive tasks, decreased mobility of the wrist & elbow, and pain that is primarily noted at the wrist. Trigger points refer pain to dorsum of the hand and index finger. Symptoms must be differentiated from those of tennis elbow, C-6 or C-7 radiculopathy, and dysfunction of serratus posterior and the scalenes.

ABDUCTOR POLLICIS LONGUS (left)
Dorsal View

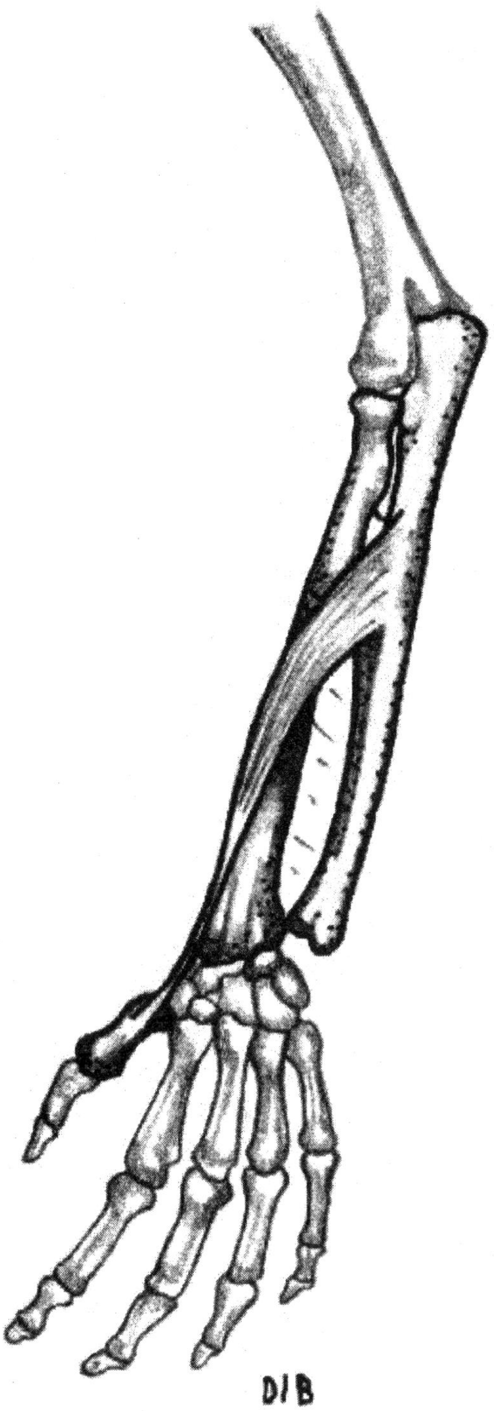

Origin:
Ulna (posterior surface of midshaft),
Radius (middle 1/3 of posterior surface of body)

Insertion:
1st metacarpal (radial side of base),
Trapezium

Action:
Abduction and extension of thumb at carpometacarpal (CMC) joint,
Radial deviation of wrist (assists),
Wrist flexion (weakly assists)

Innervation:
Radial nerve (via spinal nerves C-6 & C-7, and occasionally C-8)

Relationship:
Abductor pollicis longus lies immediately below the supinator and is occasionally fused with that muscle. It traverses obliquely downward and laterally to end in a tendon at the wrist. The tendon passes through a groove on the lateral side of the distal radius along with the tendon of the extensor pollicis brevis. Its tendon is commonly split, with one slip attached to the radial side of the 1st metacarpal and the other to the trapezium. The tendon may be palpated at the radial side of the wrist during abduction. Abductor pollicis longus may be strengthened, by abduction against resistance, from an adducted position. Symptoms associated with this muscle often include pain and tenderness in the belly of the muscle. Trigger point pain is referred from the lateral epicondyle to the web of the thumb. Complaints should be differentiated from those of carpal tunnel syndrome.

ABDUCTOR POLLICIS BREVIS (left)
Dorsal View

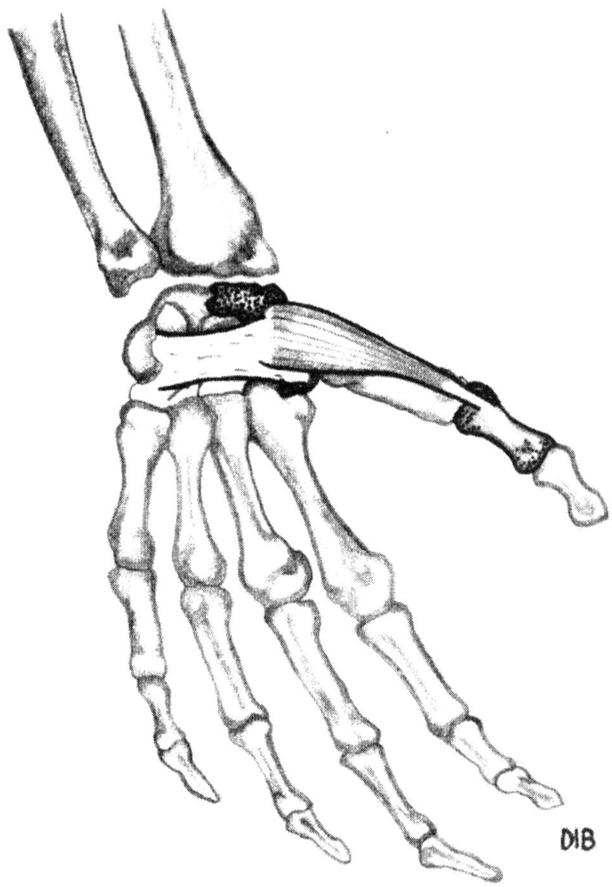

Origin:
Flexor retinaculum,
Scaphoid (tubercle),
Trapezium (tubercle)

Insertion:
Thumb (proximal phalanx, radial side of base), extensor expansion

Action:
Abduction thumb (in a plane 90° from the palm),
Opposition of thumb (assists),
Extension of IP joint (assists)

Innervation:
Median nerve (via spinal nerves C-8 & T-1)

Relationship:
Abductor pollicis brevis is a thin, superficial muscle on the radial side of the thenar eminence. It forms the bulk of muscle noted during active abduction. Deep to it are adductor pollicis and flexor pollicis brevis. Abductor pollicis brevis may be strengthened by abducting the thumb against resistance. Trigger point pain is referred to the wrist and thumb. The function of this muscle may be compromised by compression of the medial nerve. Pain and paresthesia are the most common symptoms noted. Trigger point pain projects to the radial and palmar aspects of the thumb. Complaints must be differentiated from those of carpal tunnel syndrome, tenosynovitis, and osteoarthritis.

ADDUCTOR POLLICIS (left)
Dorsal View

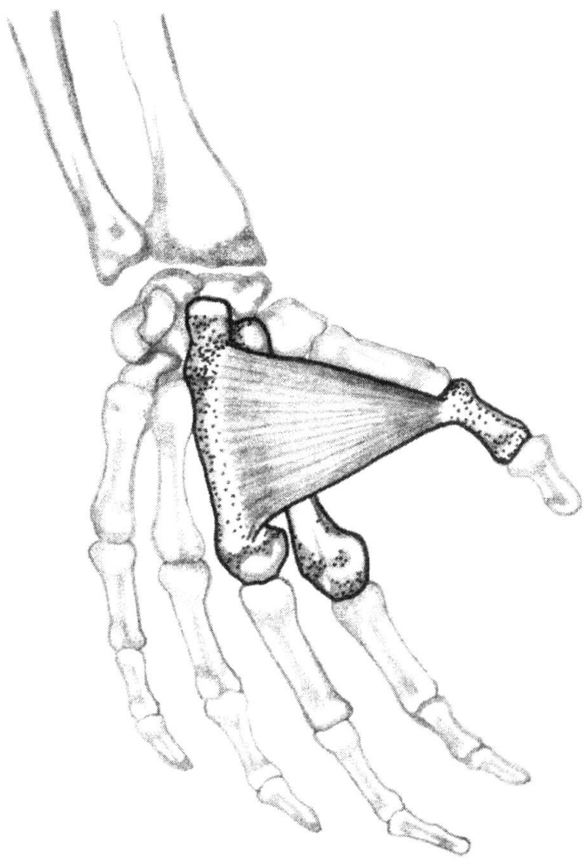

Origin:
Oblique head: Capitate, 2nd and 3rd metacarpals (bases), Palmar carpal ligaments, Flexor retinaculum (small slip), Transverse head: 3rd metacarpal (distal 2/3 of palmar surface)

Insertion (both heads):
Thumb (proximal phalanx, ulnar side of base)

Action:
Adduction of CMC joint of thumb (approximates the pollex to the palm), Adduction and flexion of MP joint (assist)

Innervation:
Ulnar nerve (via spinal nerves C-8 & T-1)

Relationship:
Adductor pollicis is a fan shaped muscle that forms the bulk of the palmar web space between the index finger and the thumb. It is easily palpated during active adduction. The fibers of the oblique head pass obliquely downward across the web of the thumb and converge into a tendon (that contains a sesamoid bone) before it unites with the transverse head. The deeper transverse head (deepest of the thenar muscles) is triangular in shape. The sizes of the two heads vary considerably. Adductor pollicis acts synergistically with flexor pollicis brevis and opponens pollicis. It may be strengthened by squeezing and gripping activities. Pain and loss of fine motor skills are the most common symptoms noted. Trigger point pain projects to the radial and palmar aspects of the thumb. Complaints must be differentiated from those of carpal tunnel syndrome, tenosynovitis, and osteoarthritis.

EXTENSOR POLLICIS LONGUS (left)
Dorsal View

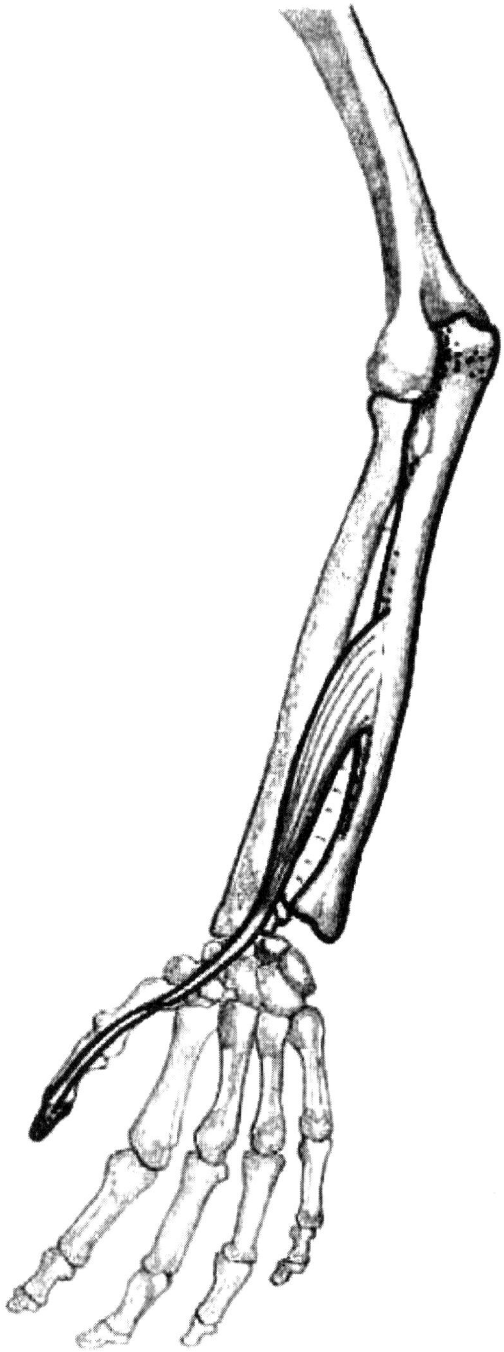

Origin:
Ulna (posterior-lateral surface of middle shaft), Interosseous membrane

Insertion:
Thumb (base of distal phalanx, dorsal surface)

Action:
Extends thumb (at IP, MP, and CMC joints), Radial deviation of wrist (accessory)

Innervation:
Radial nerve (via spinal nerves C-6 thru C-8)

Relationship:
The extensor pollicis longus arises distal to the abductor pollicis longus and courses down and laterally into a tendon. It then descends obliquely over the tendons of the carpal extensors. It can be palpated, and seen during thumb extension as the radial boundary of a triangular depression known as the *anatomical snuffbox*. Extensor pollicis longus may be strengthened by extending the flexed thumb against resistance. Trigger point pain is referred from the lateral epicondyle to the wrist and thumb. Pain and loss of fine motor skills are the most common symptoms noted. Trigger point pain is referred from the lateral epicondyle to the thumb. Complaints must be differentiated from those of lateral epicondylitis, C-7 radiculopathy, tenosynovitis, and osteoarthritis.

EXTENSOR POLLICIS BREVIS (left)

Dorsal View

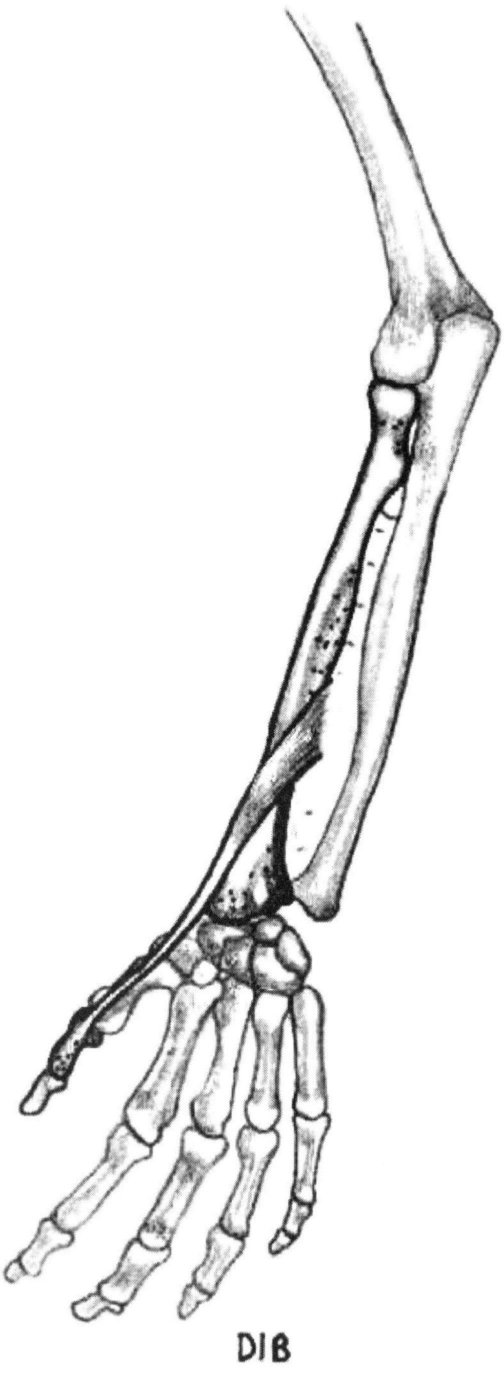

Origin:
Radius (lower portion, dorsal surface of radius),
Interosseous membrane

Insertion:
Thumb (proximal phalanx base on dorsal surface),
Attachment to distal phalanx via tendon of extensor pollicis longus is common

Action:
Extension of MP joint of thumb extension,
Abduction of 1st CMC joint of thumb,
Radial deviation of wrist (accessory)

Innervation:
Radial nerve (via spinal nerves C-6, C-7 and occasionally C-8)

Relationship:
Extensor pollicis brevis arises distal to and lies medial to the abductor pollicis longus and descends with it so that the tendons of the two muscles pass through the same groove on the lateral side of the distal radius. The muscle often is actually connected with the abductor. It also may be absent in some cases. When present, it can be palpated just lateral to the tendon of extensor pollicis longus. Symptoms associated with this muscle often include (TP) pain and tenderness in the belly of the muscle that refers to wrist and thumb. Complaints must be differentiated from those of lateral epicondylitis, C-7 radiculopathy, tenosynovitis, and osteoarthritis.

FLEXOR POLLICIS BREVIS (left)
Ventral (Palmar) View

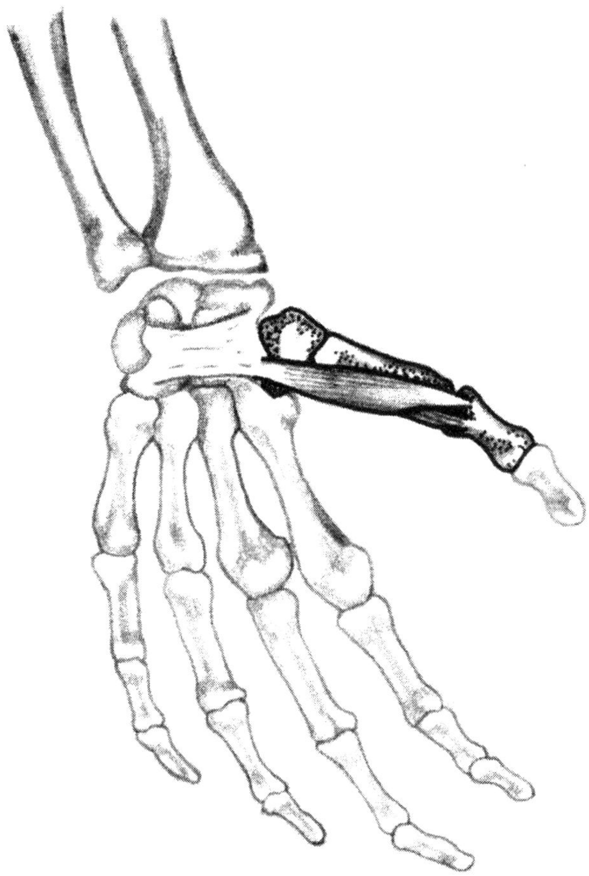

Origin:
Superficial head: flexor retinaculum (distal border),
Trapezium (tubercle),
Deep head: trapezoid, capitate

Insertion:
Thumb (base of proximal phalanx on radial side),
Extensor expansion

Action:
Flexion of the MP and CMC joints of the thumb,
Opposition of thumb (assist)

Innervation:
Median nerve, (C-8, T-1 superficial head)
Ulnar nerve, (C-8, T-1 deep head)

Relationship:
The superficial head runs more laterally and accompanies the flexor pollicis longus. Its tendon of insertion contains the radial sesamoid bone at a point where it unites with the tendon of the deep head. The deep head is sometimes absent. Flexor pollicis brevis acts with opponens pollicis, flexor pollicis longus, and adductor pollicis in gripping and squeezing activities. This muscle can be palpated on the palmar surface of the proximal phalanx during active flexion. Pain and loss of fine motor skills are the most common symptoms noted. Trigger point pain is referred to the wrist and thumb. Complaints must be differentiated from those of carpal tunnel syndrome, tenosynovitis, and osteoarthritis.

OPPONENS POLLICIS (left)
Ventral (Palmar) View

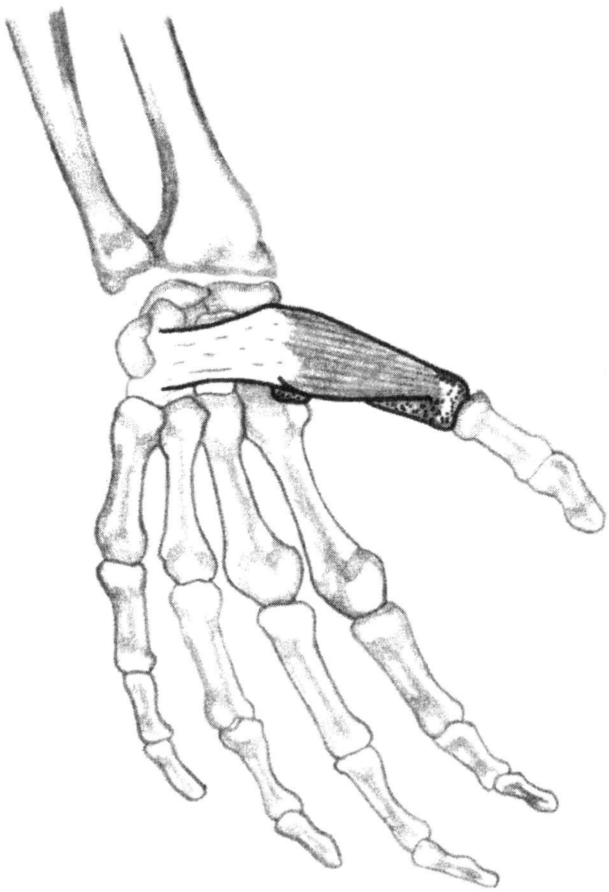

Origin:
Trapezium (tubercle),
Flexor retinaculum

Insertion:
1st metacarpal (along entire length of radial side of shaft)

Action:
Flexion of CMC joint medially across the palm,
Abduction of CMC joint,
Medial rotation of CMC joint
These motions occur simultaneously in the motion called opposition, which brings the thumb into contact with any of the other fingers on their palmar digital aspect (pads)

Innervation:
Median nerve (via spinal nerves C-8, T-1)

Relationship:
Opponens pollicis is a small triangular muscle, lying deep to the abductor pollicis. Abductor pollicis brevis, flexor pollicis brevis, adductor pollicis, and opponens pollicis comprise the thenar eminence (muscular pad) on the palmar surface of the 1st metacarpal. It may be strengthened by squeezing and gripping activities. Pain and loss of fine motor skills are the most common symptoms noted. Trigger point pain is referred to the wrist and thumb. Complaints must be differentiated from those of carpal tunnel syndrome, tenosynovitis, and osteoarthritis.

OPPONENS DIGITI MINIMI (left)
Ventral (Palmar) View

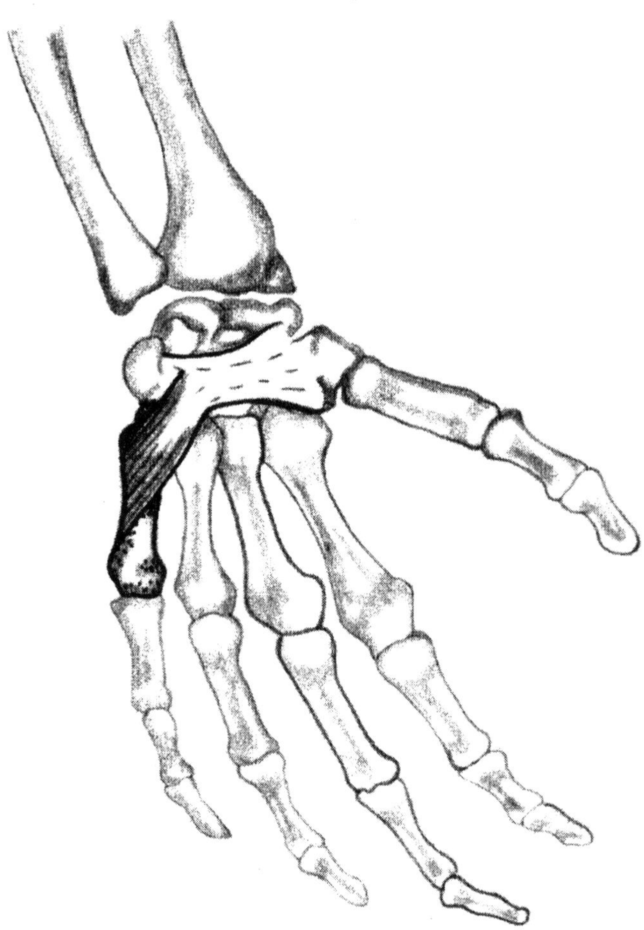

Origin:
Hook of hamate, and adjacent portion of flexor retinaculum

Insertion:
5th metacarpal (entire length of ulnar margin)

Action:
Opposition of little finger (abduction, flexion and lateral rotation, deepening the palmar hollow) to the thumb

Innervation:
Ulnar nerve (via spinal nervres C-8, T-1)

Relationship:
Opponens digiti minimi is a triangular muscle found deep to the abductor and the flexor. It is commonly blended with its neighbors and is difficult to palpate. It may be strengthened by squeezing and gripping activities. Pain, loss of fine motor skills, and weakness are the most common symptoms noted. Trigger point pain is referred to the wrist and little finger. Complaints must be differentiated from those of ulnar nerve entrapment, tenosynovitis, and osteoarthritis.

ABDUCTOR DIGITI MINIMI (left)
Ventral (Palmar) View

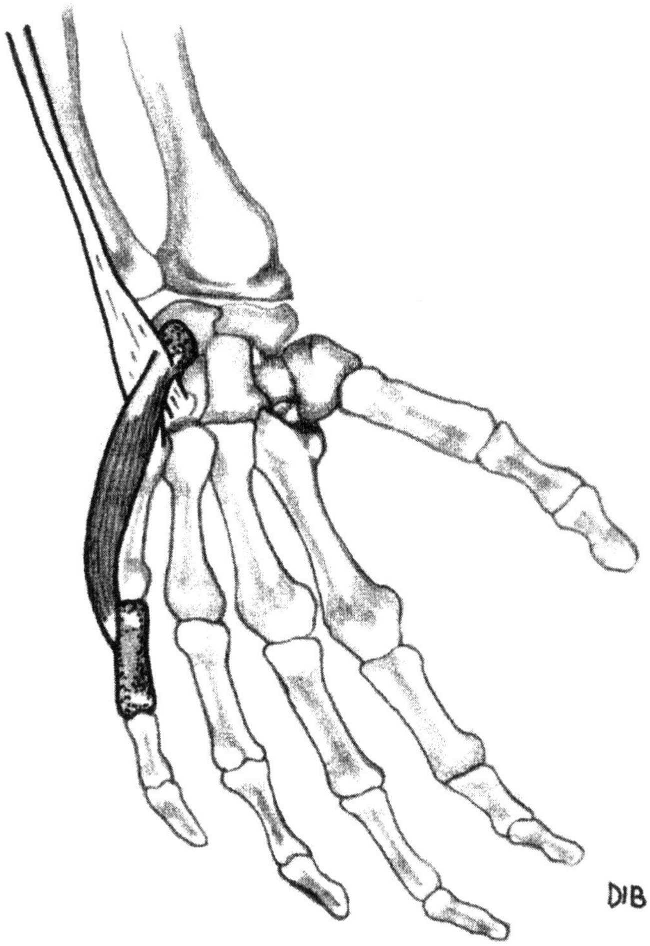

Origin:
Pisiform bone,
Tendon of flexor carpi ulnaris

Insertion:
5th digit (base of proximal phalanx on ulnar aspect),
Into dorsal digital expansion of extensor digiti minimi

Function:
Abduction of 5th digit away from ring finger,
Flexion of proximal phalanx of 5th digit at the MP joint,
Opposition of 5th digit (assist)

Innervation:
Ulnar nerve (via spinal nerves C-8, T-1)

Relationship:
Abductor digiti minimi can be palpated on the dorsal aspect of the 5th metacarpal during active abduction. It contributes to the formation of the hypothenar eminence. It may be strengthened by squeezing and gripping activities. Pain and loss of fine motor skills are the most common symptoms noted. Trigger point pain is referred to the wrist and little finger. Complaints must be differentiated from those of ulnar nerve entrapment, tenosynovitis, and osteoarthritis.

DORSAL INTEROSSEI (right)

Dorsal View

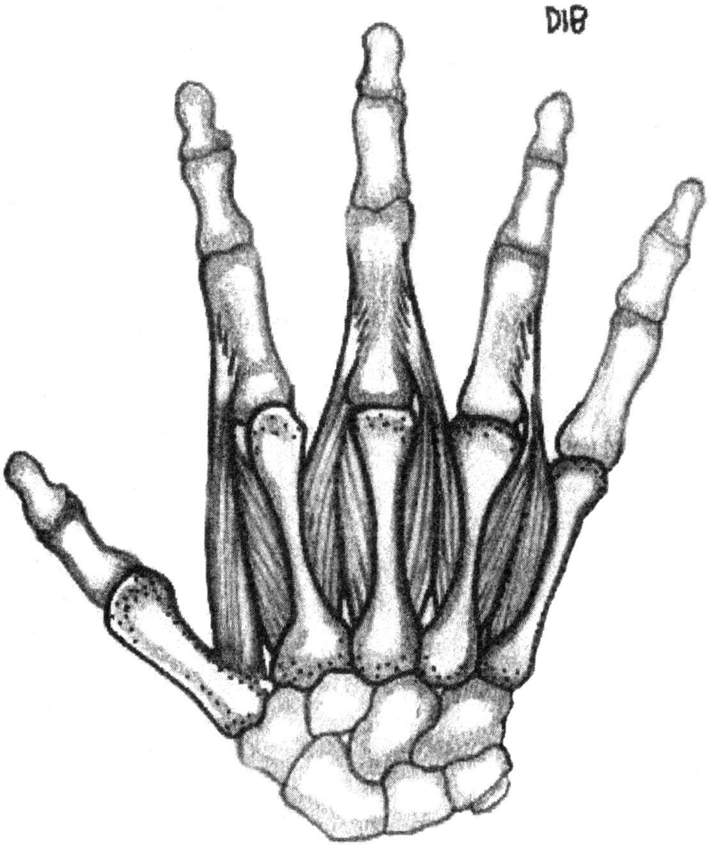

Origin:
Shafts of adjacent metacarpal bones (dorsal aspect), via two heads

Insertion:
Bases of the 2nd, 3rd, and 4th proximal phalanges and their extensor expansions

Action:
Abduction of fingers away from an axis drawn through the center of the middle finger (**D**orsal Interossei **Ab**duct = **DAB**)

Flexion of fingers at MP joints (assist),
Extension of fingers at IP joints (assist),
Thumb adduction (assist)

Innervation:
Ulnar nerve, palmar branch (via spinal nerves C-8 & T-1)

Relationship:
This group consists of four bipennate muscles. They can be palpated along the dorsal and lateral aspects of the 2nd, 3rd, and 4th metacarpals during active abduction. Pain, paresthesia, and the loss of fine motor skills, such as writing, are the most common symptoms noted. Trigger point pain is commonly referred to the distal joints. Complaints must be differentiated from those of thoracic outlet syndrome, ulnar nerve entrapment, tenosynovitis, and osteoarthritis.

PALMAR INTEROSSEI (right)
(shown with abductor digiti minimi)
Palmar View

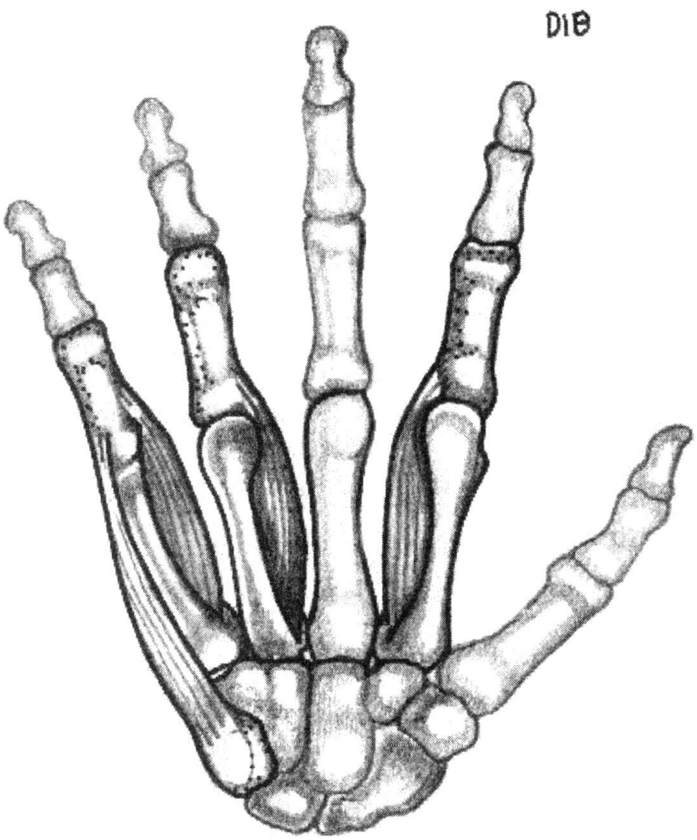

Origin:
Shafts of 2nd, 4th, and 5th metacarpals and their extensor expansions

Insertion:
Bases of the 2nd, 4th, and 5th proximal phalanges and their extensor expansions

Action:
Adduction of fingers (index, ring, and little) toward an axis drawn through the center of the middle finger,
(**P**almar Interossei **Ad**duct = **PAD**)
Flexion of MP joints (assist),
Extension of IP joints (assist)

Innervation:
Ulnar nerve (via spinal nerves C-8 & T-1)

Relationship:
The palmar interossei are smaller than their dorsal counterparts. They are found on the palmar surface of the hand at the metacarpal bones There are three very distinct volar interossei. The middle finger has no interosseous muscle. Pain, paresthesia, and the loss of fine motor skills, such as writing, are the most common symptoms noted. Trigger point pain is commonly referred to the distal joints. Complaints must be differentiated from those of thoracic outlet syndrome, ulnar nerve entrapment, tenosynovitis, and osteoarthritis.

LUMBRICALES (left)
Palmar View

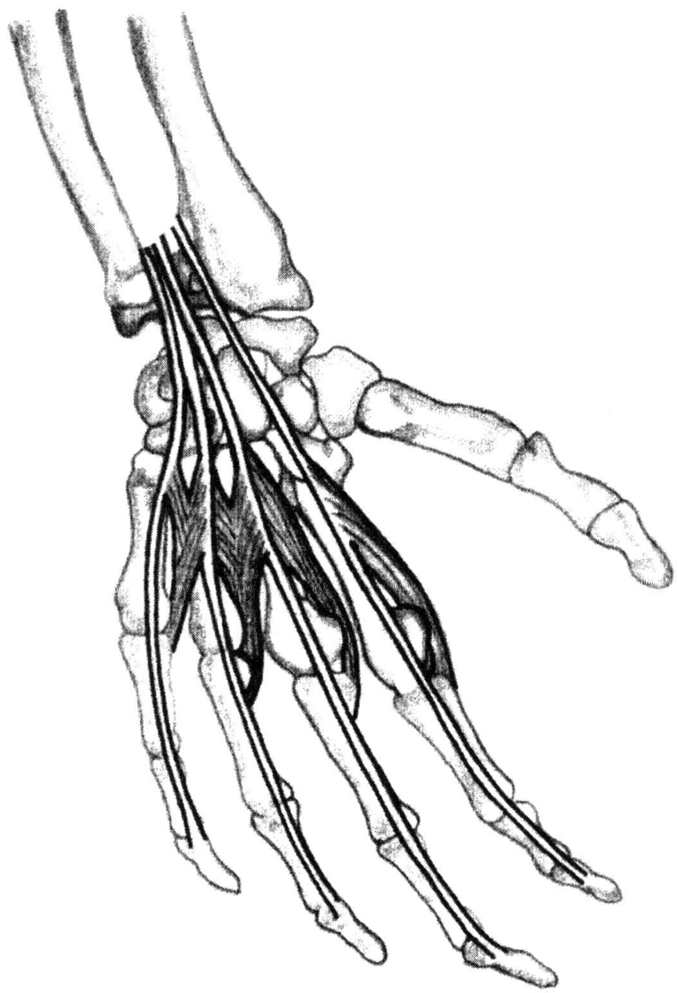

Origin:
Flexor digitorum profundus tendons (near the center of the palm)

Insertion:
Extensor expansions on the radial aspect of 2nd, 3rd, 4th, and 5th proximal phalanges

Action:
Flexion of MP joints (proximal phalanges) of digits 2 to 5 and simultaneous extension of the IP joint

Innervation:
1st and 2nd - Median nerve (via C-8 & T-1)
3rd and 4th - Ulnar nerve (via C-8 & T-1)

Relationship:
These four, small muscles are odd in that they arise from the tendons of the flexor digitorum profundus as they pass the metacarpals. They may be unipennate or bipennate. They extend to the middle phalanges of digits 2 to 5, where they join the dorsal extensor expansion on the radial side of each digit. This arrangement links the flexor to the extension tendon systems in the hand, and gives rise to great complexity of movement. This flexion-extension function is demonstrated when strongly gripping an object without using fingertip pressure. Pain, paresthesia, and the loss of fine motor skills, such as writing, are the most common symptoms noted. Trigger point pain is commonly referred to the distal joints. Complaints must be differentiated from those of thoracic outlet syndrome, ulnar nerve entrapment, tenosynovitis, and osteoarthritis.

INTRINSIC MUSCLES of the HAND (right)

Anterior (Palmar) View

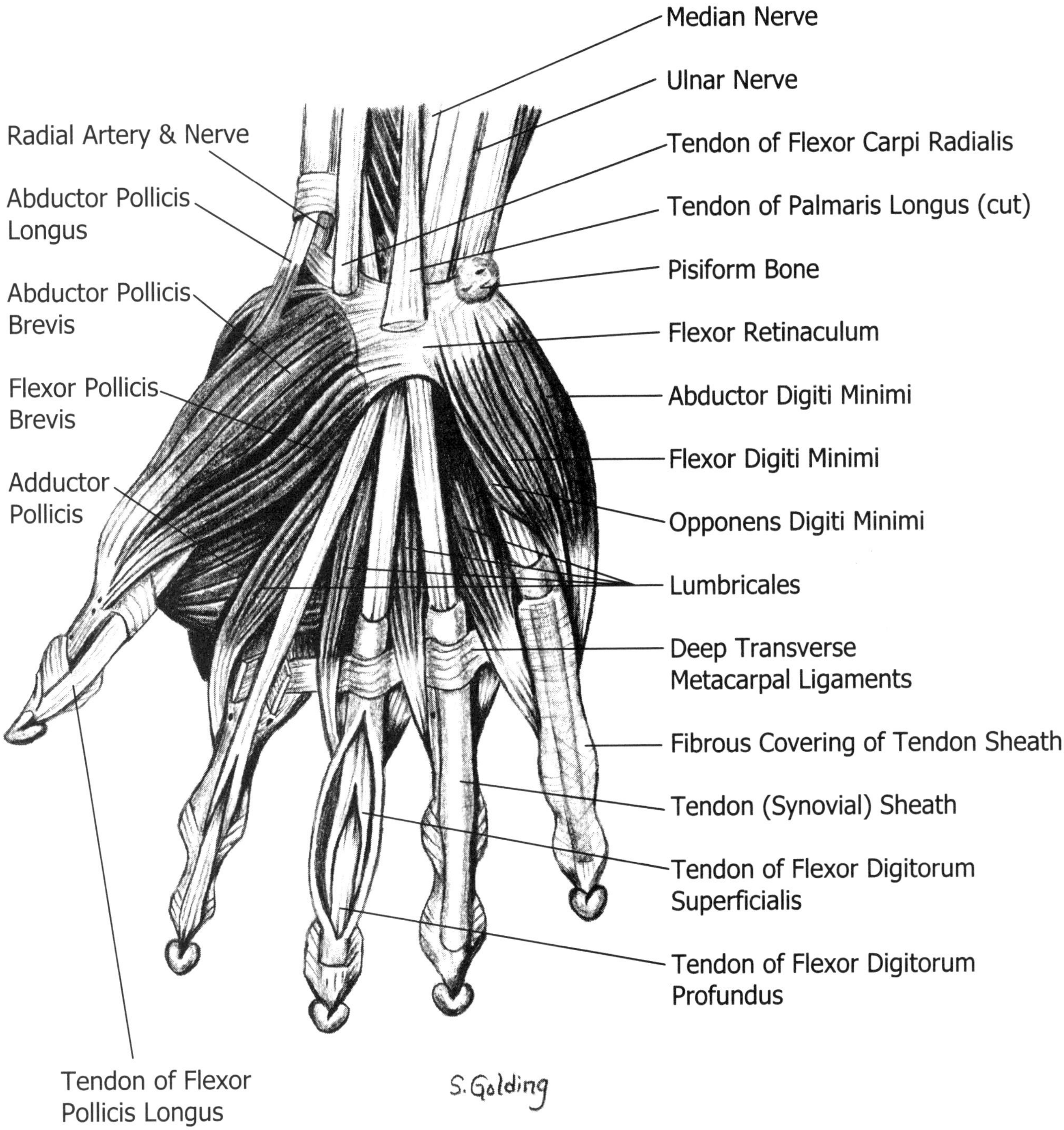

Iliopsoas (right)
Anterior View

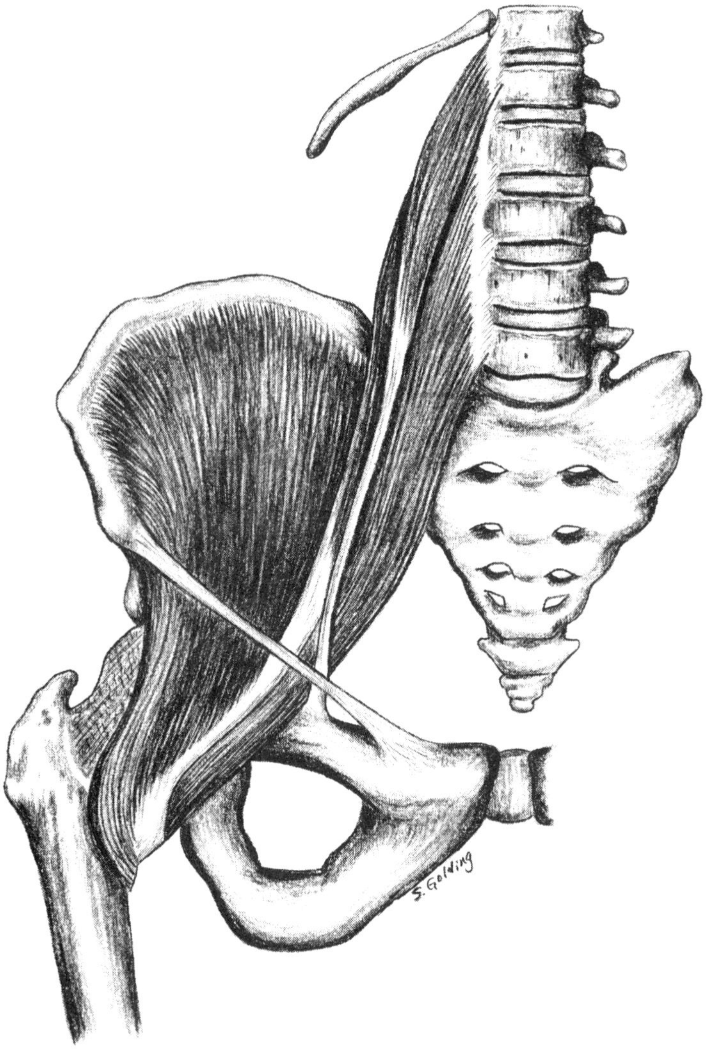

The Iliopsoas is a complex muscle formed by the iliacus, psoas major, and psoas minor muscles. Its superior anatomical attachments include the T-12 thru L-5 vertebra and discs, the anterior surfaces of the ilium, and the anterior sacroiliac and iliolumbar ligaments. Its inferior attachments are primarily the lesser trochanter, adjacent areas of the femur, the iliopectineal eminence, and pectineal line of the ilium. The Iliopsoas is covered anteriorly by the iliac fascia and shares space inferiorly with the femoral nerve, femoral artery, and numerous lymphatic structures. Although presented here as a single muscle, this muscle is commonly described as being three distinct muscles. The iliopsoas functions to flex the thigh at the hip and to assist in extension of the lumbar spine (increasing lumbar lordosis). It is active in sitting, standing, walking, and running. It is very active during the latter portion (60°) of doing a strait-leg sit-up. It is thought to be involved in the development of scoliosis. Iliopsoas is synergistic (in hip flexion) with rectus femoris and pectineus. It is commonly involved in cases of low back pain, which is made worse by standing and better by lying. Iliopsoas may also present with symptoms of lower right quadrant pain, upper thigh pain and constipation. Active iliopsoas trigger points may refer pain to the mid/low back region, anterior thigh, and groin region. Complaints must be differentiated from those of quadratus lumborum, the adductors, tensor fascia latae, and acute appendicitis. The iliopsoas is stretched by extending the hip far enough to move the femur to a position posterior to the frontal plane of the body.

PSOAS MAJOR (left)

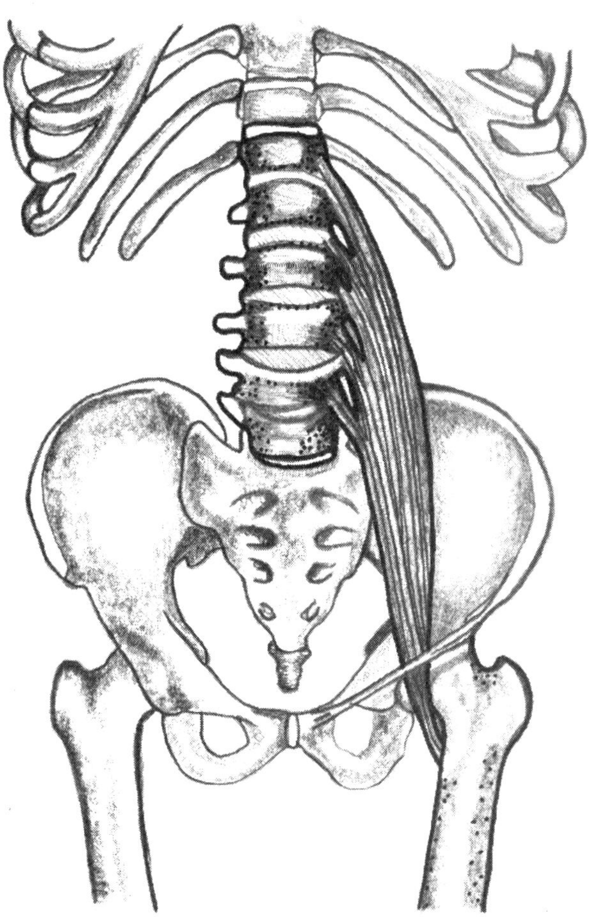

Origin:
T-12 thru L-5 vertebral bodies and intervertebral discs between them (via tendinous faciculi), L-1 thru L-5 vertebrae, (via tendinous faciculi to the inferior borders of transverse processes)

Insertion:
Lesser trochanter of femur (via iliopsoas tendon)

Action:
Hip flexion (with origin fixed),
Trunk flexion (sit-up) with insertion fixed (these two functions occurring in conjunction with the Iliacus),
External (lateral) rotation of femur at hip (slight),
Unilateral: lateral bending of lumbar spine to ipsilateral side (slight)

Innervation:
Lumbar plexus (via fibers from spinal nerves L-2 thru L-4, ventral rami)

Relationship:
Psoas major lies just anterior to the transverse processes of the lumbar spine. It passes inferiorly and immediately anterior to the sacroiliac and the hip joints to join with the iliacus and form the iliopsoas. The psoas is active in sitting, standing, walking, and running. It has been shown to be very active during the latter portion (after 30º) of a sit-up. Psoas is synergistic (in hip flexion) with rectus femoris and pectineus. Psoas is best strengthened by leg raises from a dip bar. It is commonly involved in cases of low back pain, which is made worse by standing and better by lying, and is thought to be involved in the development of scoliosis. Psoas may also present with symptoms of lower right quadrant pain, upper thigh pain, and constipation. Trigger points refer to the low back, hip, and groin. Complaints must be differentiated from those of quadratus lumborum, the adductors, and tensor fascia latae.

PSOAS MINOR (left)

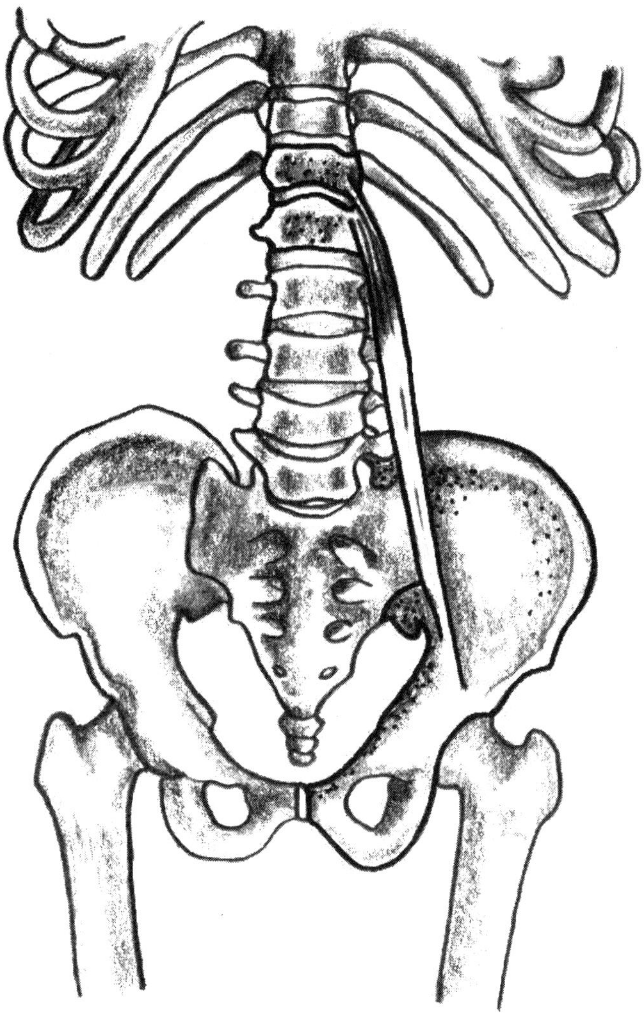

Origin:
T-12 & L-1 vertebral bodies (anterior & lateral portions) and the intervertebral discs between them

Insertion:
Ilium (iliopectineal eminence and linea terminalis or pectineal line of inner surface of the pelvis),
Iliac fascia

Action:
Flexion of trunk and lumbar spine (both weakly)

Innervation:
L-1 spinal nerve (branch)

Relationship:
This muscle lies anterior to the psoas major. It is a long, slender muscle whose belly lies entirely within the abdominopelvic region and along its posterior wall. The muscle is present in only 50% to 60% of specimens. It is active during sit-ups and could likely contribute to scoliosis. Psoas minor symptoms (right sided) are commonly mistaken for those of acute appendicitis.

ILIACUS (left)

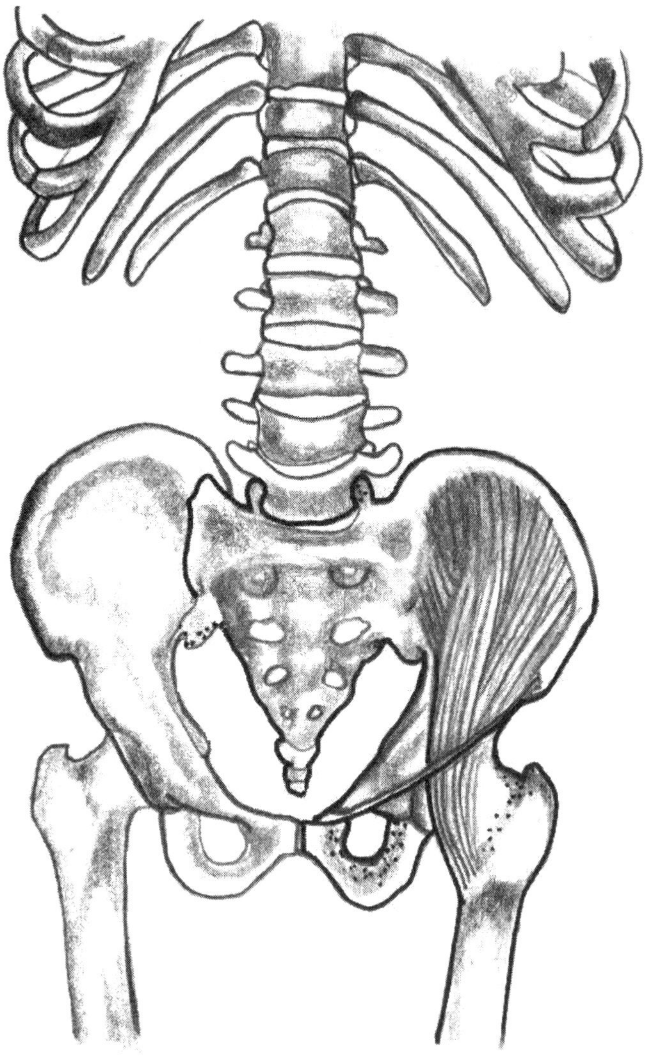

Origin:
Ilium (superior 2/3 of iliac fossa),
Iliac crest (inner lip),
Anterior sacroiliac and iliolumbar ligaments

Insertion:
Lesser trochanter of femur (via iliopsoas tendon) and shaft adjacent to the lesser trochanter

Action:
Hip flexion,
Trunk flexion (occurring only in conjunction with psoas major),
External rotation of femur at hip

Innervation:
Femoral nerve (via spinal nerves L-2 & L-3)

Relationship:
Iliacus is a broad, flat muscle, filling the iliac fossa and descending along the fossa, converging laterally to join the tendon of the psoas major. Iliacus crosses only the hip joint. The large iliopectineal bursa lies between this muscle and the capsule of the hip joint, while the subtendinous bursa separates the iliopsoas from the femur at its attachment. While the iliacus is relatively quiet during standing, it acts continuously during walking and running. Like psoas major, iliacus becomes very active during the latter portion (after 30°) of a sit-up. Leg raises from a dip or parallel bar strengthen this muscle. Clinical considerations are similar to those of psoas major.

RECTUS FEMORIS (left)

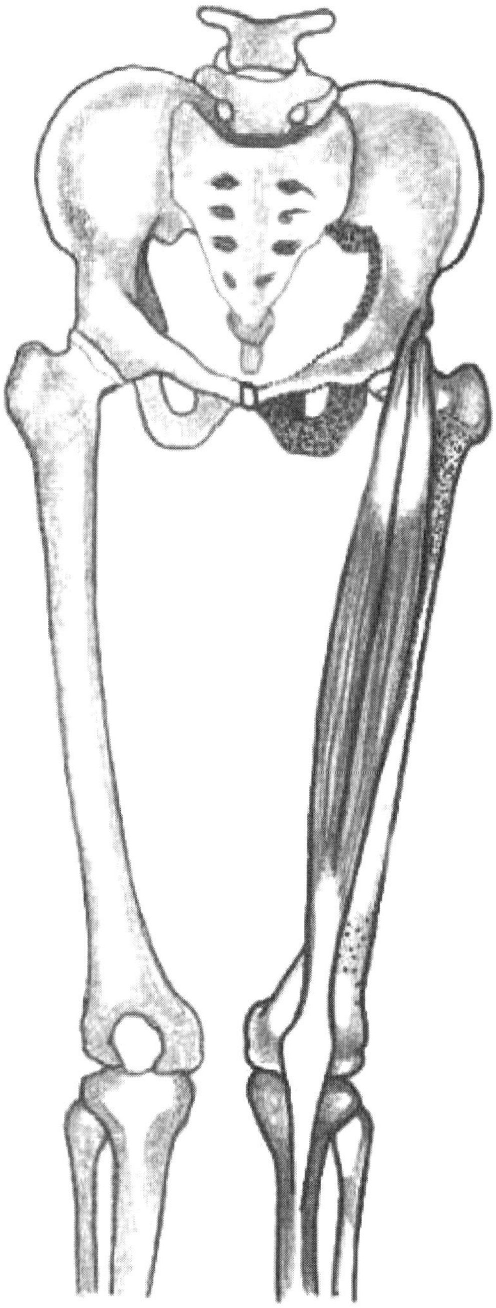

Origin:
Ilium, from two tendons, which conjoin to form an aponeurosis from which the muscle fibers arise, Anterior inferior iliac spine & acetabulum (groove above posterior rim and superior margin of labrum)

Insertion:
Patella (from an aponeurosis that gradually narrows into a tendon which forms the central portion of the quadriceps tendon and thus onward via the patellar ligament to the tibial tuberosity)

Action:
Knee extension, (in conjunction with the vasti)
Hip flexion (flexes the femur at the hip)

Innervation:
Femoral nerve (via spinal nerves L-2 thru L-4)

Relationship:
Rectus femoris is the most anterior of the quadriceps, lying 6º medial to the axis of the femur. Its superficial fibers are bipinate, but the deep fibers are parallel. Rectus femoris traverses the length of the femur and is covered proximally by sartorius. This muscle exerts a proximal pull on the patella and balances tension from the vasti, to ensure normal patellar tracking. Although nearly at rest in the standing position and when rising, rectus becomes prominent during high-speed movements. It is very active in heavy lifting and jumping. Rectus is less active than the vasti in the down stroke of cycling. It is strengthened by hip flexion and knee extension against resistance. Common symptoms include anterior knee pain (worse at night), referred pain and weakness & pain when descending stairs. Rectus femoris may be aggravated by trauma, prolonged sitting (without stretching), and by chronic overload due to faulty biomechanics. Trigger point pain is referred to the distal anterior thigh and knee. Symptoms must be differentiated from those of tendonitis, bursitis, sprain/strain, joint dysfunction, meniscus tear, and chondromalacia patellae.

VASTUS INTERMEDIUS (left)

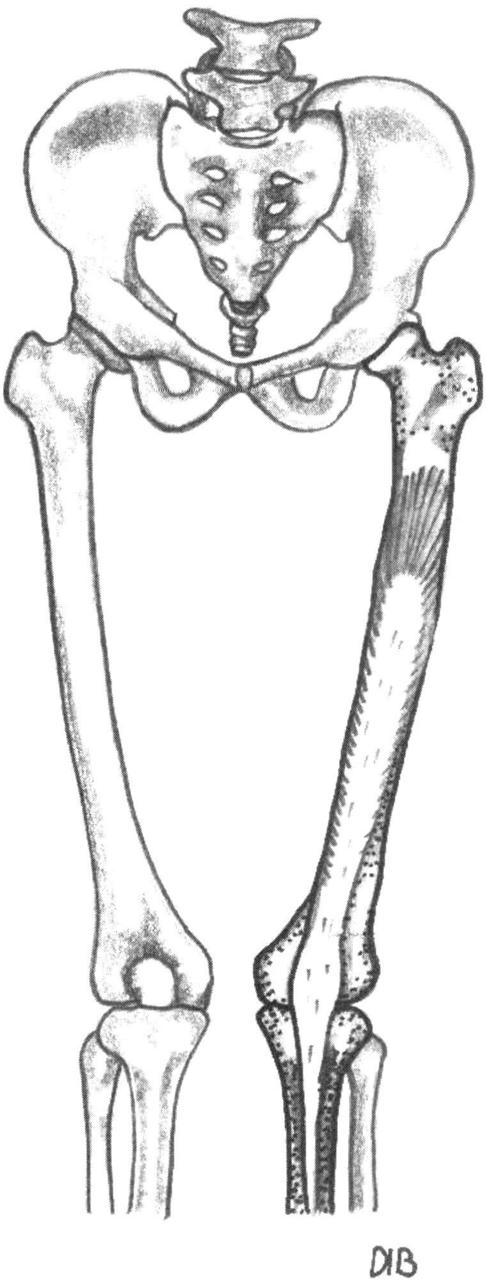

Origin:
Femur (anterior and lateral surfaces of upper 2/3 of shaft),
Lateral intermuscular septum (lower part)

Insertion:
Patella (from an aponeurosis on the anterior surface of the muscle, which attaches to the middle part of the quadriceps tendon and thus onward via the patellar ligament to the tibial tuberosity)

Action:
Knee extension (in conjunction with the other vasti)

Innervation:
Femoral nerve (via spinal nerves L-2 thru L-4)

Description:
Vastus intermedius lies deep to rectus femoris, vastus medialis, and vastus lateralis. It almost completely surrounds the proximal 2/3 of the shaft of the femur and its lateral fibers merge with those of vastus lateralis. The articularis genus is occasionally distinguishable from the intermedius, but more commonly, it is a part of the intermedius. When present, the articularis genus functions to retract the supra-patellar bursa during extension of the knee. Vastus intermedius is active in walking, heavy lifting, jumping, cycling (down stroke), and especially in stair climbing. It is strengthened by knee extension exercises against resistance. Associated complaints often include an inability to straighten the knee (worse after sitting) and pain upon climbing stairs. Trigger point pain is referred to the middle portion of the anterior thigh. Symptoms must be differentiated from those of tendonitis, bursitis, sprain/strain, joint dysfunction, meniscus tear, and chondromalacia patellae.

VASTUS MEDIALIS (left)

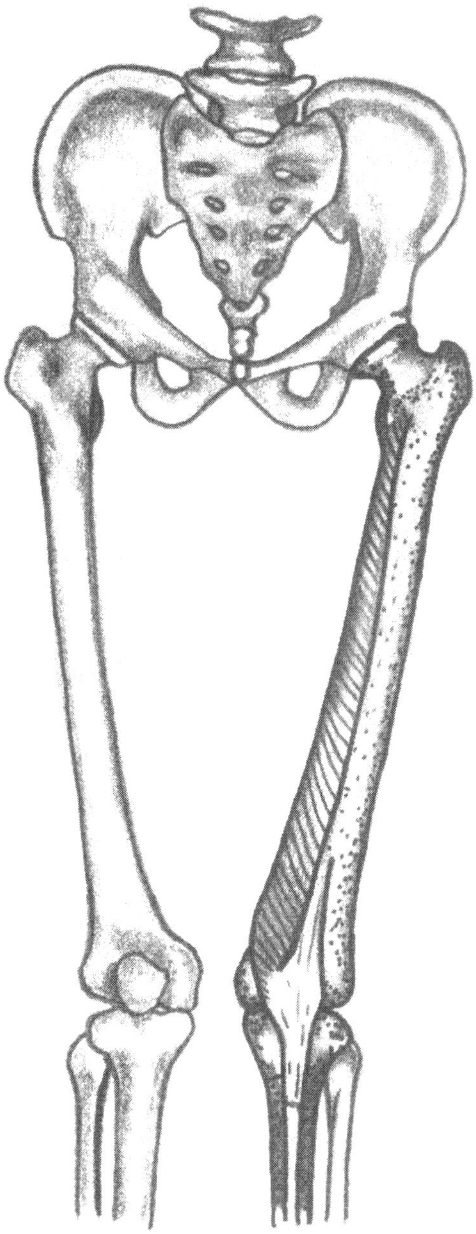

Origin:
Femur (proximal portion of posteromedial aspect, intertraochanteric line, medial lip of linea aspera), tendons of adductors longus and magnus and the medial intermuscular septum

Insertion:
Patella (via an aponeurosis into the superior medial margin of the quadriceps tendon and thus onward via the patellar ligament to the tibial tuberosity) and to the medial patellar retinaculum

Action: Knee extension (in conjunction with the other vasti). Some authors attribute the last 15º of extension primarily to vastus medialis

Innervation:
Femoral nerve (via spinal nerves L-2 thru L-4)

Relationship:
The majority of fibers of this muscle course upward at an angle of 15º to 18º to the longitudinal axis of the femur. The distal angulated fibers of this muscle are at a more oblique angle of 50º to 55º and are designated as the "vastus medialis oblique". This portion of the muscle appears to bulge quickly with training and to atrophy with disuse before the other quadriceps show changes. This phenomenon is explained by the fact that the medialis oblique has the most sparse and thinnest fascial investment, making changes in it more obvious to observation. Vastus medialis is the dominate muscle used in walking with a load, heavy lifting, running, jumping, cycling (down stroke), rising from a seated position, and stair climbing. Symptoms from vastus medialis often include a deep toothache-like pain (which is worse during sleep), referred pain, and weakness (buckling) of the knee while walking. All symptoms would be made worse by excessive pronation (flat feet) and strenuous athletic activity. Trigger point pain is referred to the anteriomedial aspect of the thigh and knee. Symptoms must be differentiated from those of tendonitis, bursitis, sprain/strain, joint dysfunction, meniscus tear, and chondromalacia patellae.

VASTUS LATERALIS (left)

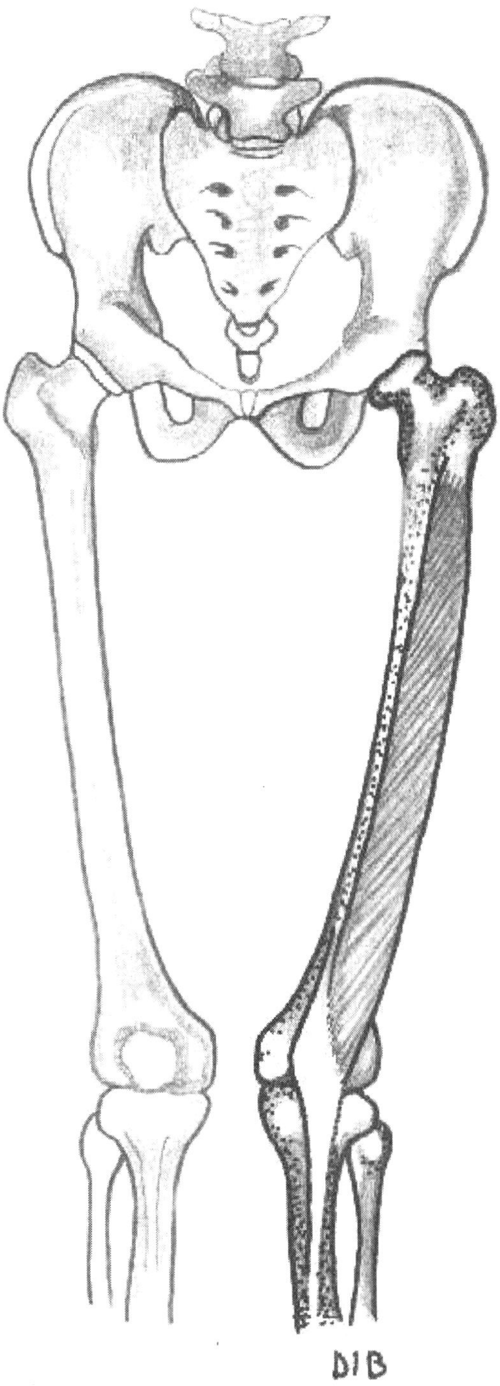

Origin:
Femur (lateral portion of the posterior aspect of the proximal ¾ via an aponeurosis arising deep to the muscle)

Insertion:
Patella (via an aponeurosis into the superior lateral margin of the quadriceps tendon and thus onward via the patellar ligament to the tibial tuberosity) and to the lateral patellar retinaculum

Action:
Knee extension (in conjunction with the other vasti)

Innervation:
Femoral nerve (via spinal nerves L-2 thru L-4)

Relationship:
This is the largest component of the quadriceps femoris and forms the bulk of the lateral thigh musculature. Its fibers run at an angle of 17° to the axis of the femur. It is deep to the iliotibial band as it descends the thigh. Vastus lateralis is less used in walking, lifting, running, jumping, and cycling, yet is more active in checkreining knee flexion with eccentric (lengthening) contractions. Symptoms from vastus lateralis often include lateral thigh and knee pain (made worse by lying on involved side), referred pain, and locking of the patella (which usually immobilizes the knee in a slightly flexed position). It is strengthened by knee extension exercises against resistance. Trigger point pain may be referred along the lateral aspect of the upper thigh, down to the knee. Symptoms must be differentiated from those of tendonitis, bursitis, sprain/strain, joint dysfunction, meniscus tear, and chondromalacia patellae.

SARTORIUS (left)

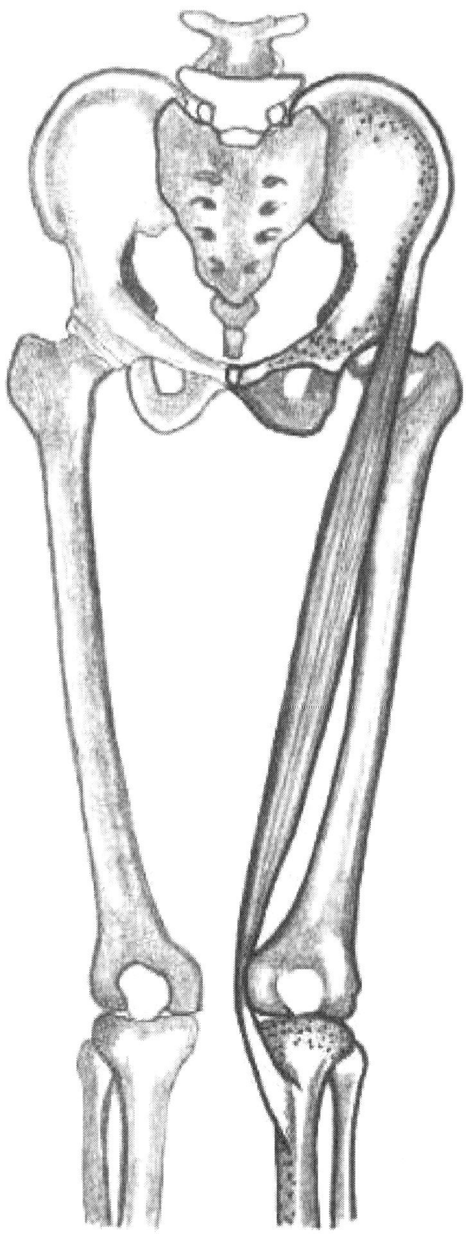

Origin:
Ilium (anterior superior iliac spine and notch below the ASIS)

Insertion:
Tibia (medial surface of the shaft distal to the tibial condyle, anterior to the attachments of gracilis & semitendinosus)

Action:
Hip flexion, with external rotation and abduction,
Knee flexion and internal rotation (slight)

Innervation:
Femoral nerve (usually two branches, via spinal nerves L-2 & L-3)

Relationship:
Sartorius is the longest muscle in the body and is the most superficial of the anterior thigh muscles. Its parallel fibers, form a narrow, thin strap that descend obliquely from lateral to medial to just above the knee, where it turns abruptly downward and passes posterior to the medial condyle of the femur. Sartorius expands into a broad aponeurosis before inserting on the medial surface of the tibia. At its distal attachment, sartorius is the most medial of the pes anserinus muscles. Its name is derived from assisting the movements necessary to assume the position (cross-legged) of a tailor (*sartor* is the Latin word for tailor). Sartorius is active in walking, jumping, tennis, throwing, batting, and cycling. It is best strengthened by leg raises from a dip or parallel bar. Symptoms of sartorius commonly include superficial pain in the anterior thigh & knee and paresthesia of the anteriomedial thigh. It may also be involved in an entrapment of the lateral femoral cutaneous nerve. Trigger points produce sharp, tingling, superficial, and localized bursts of pain. Symptoms must be differentiated from those of vastus medialis and knee related tendonitis, bursitis, sprain/strain, arthritis, or joint dysfunction.

PECTINEUS (left)
Anterior View

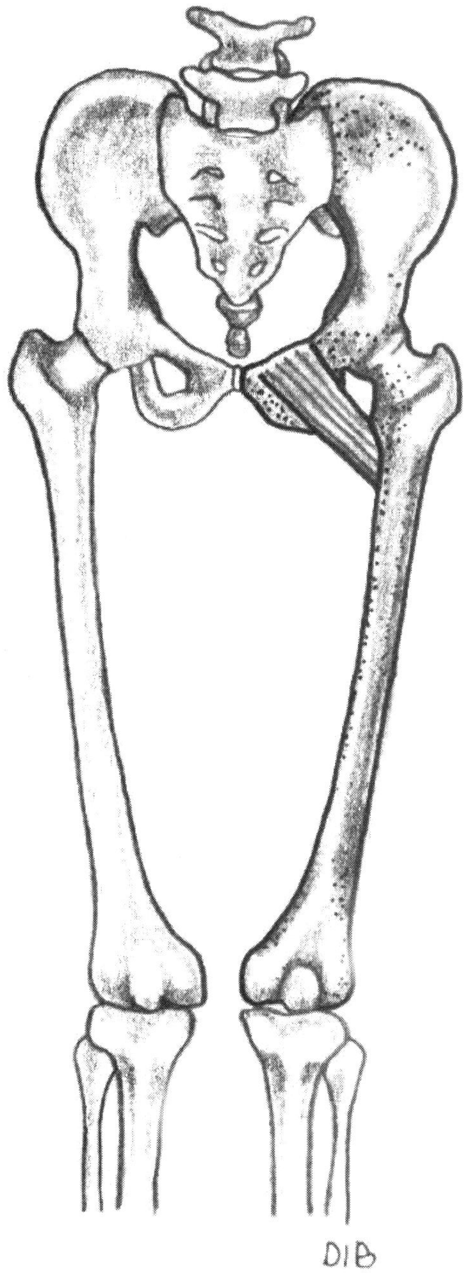

Origin:
Pubis (pecten or crest of superior ramus, lateral to the pubic tubercle)

Insertion:
Femur (pectineal line on medial, posterior surface)

Action:
Hip adduction
Hip flexion (accessory)

Innervation:
Femoral nerve & accessory obturator nerve when present (via spinal nerves L-2 thru L-4)

Relationship:
Pectineus is a flat muscle forming most of the medial portion of the femoral triangle in the upper medial aspect of the thigh. It is just deep to the neurovascular bundle containing the large femoral artery and nerve. It descends posteriorly and laterally on the medial thigh to its distal attachment. Pectineus is synergistic with gracilis, the adductors (longus, brevis, & magnus), and iliopsoas. It is opposed in adduction by gluteus medius & minimus, and the tensor fascia latae. It is opposed in flexion by gluteus maximus and the hamstrings. Pectineus is active in horseback riding, gymnastic exercises, and heavy lifting. It is strengthened by hip flexion and hip adduction exercises against resistance. Symptoms are aggravated by overuse and commonly include referred pain, deep groin pain, and decreased ROM in abduction. Trigger point pain may be referred from the inguinal ligament, down the anteriomedial thigh. Symptoms must be differentiated from those of obturator nerve entrapment, pubic stress symphysitis, and hip joint disorders (arthritis, pelvic subluxation/tilt).

ADDUCTOR LONGUS (right)
Anterior View

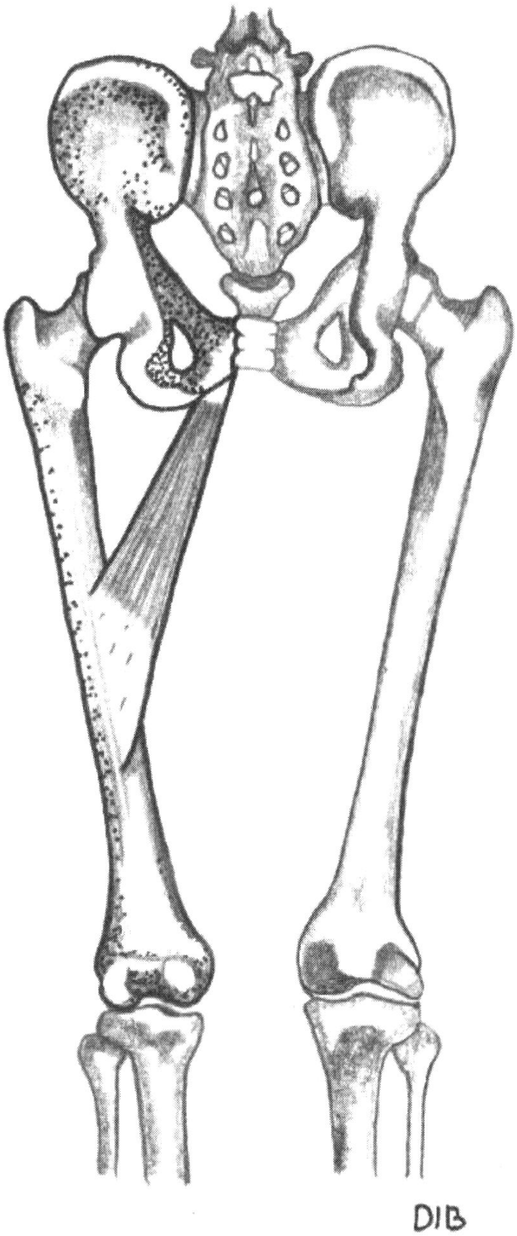

Origin:
Pubis (via a flat tendon to an area on the outer surface of the pelvis between the obturator foramen and the symphysis pubis)

Insertion:
Femur (by an aponeurosis on the middle 1/3 of the linea aspera on its medial lip)

Action:
Adduction of femur at hip,
Flexion of femur at hip (accessory),
Internal rotation of femur (accessory)

Innervation:
Obturator nerve (anterior division, via spinal nerves L-2 thru L-4)

Relationship:
Adductor longus is the most anterior of the adductor muscles. It arises from a narrow tendon and widens into a broad muscle belly as it descends posteriorly and laterally to insert upon the femur. Adductor longus is synergistic in adduction with brevis, magnus, pectineus, and gracilis. It is opposed in that action by gluteus medius & minimus, and tensor fascia latae. In internal (medial) rotation, the adductor group acts with the anterior portion of gluteus minimus, and is opposed by the deep lateral rotators and gluteus maximus. Adductor longus is active in walking, running, jumping, stair climbing, and horseback riding. It is strengthened by adduction against resistance. Symptoms are aggravated by strenuous activity, overuse, and commonly include knee, thigh, groin pain (worse with weight bearing & twisting of hip), and decreased ROM in abduction & lateral rotation. Trigger point pain may be referred upward to deep in the groin and downward to the knee and shin. Symptoms must be differentiated from those of vastus medialis & pectineus dysfunction, lumbar disc disease, and nerve entrapment.

ADDUCTOR BREVIS (right)
Anterior View

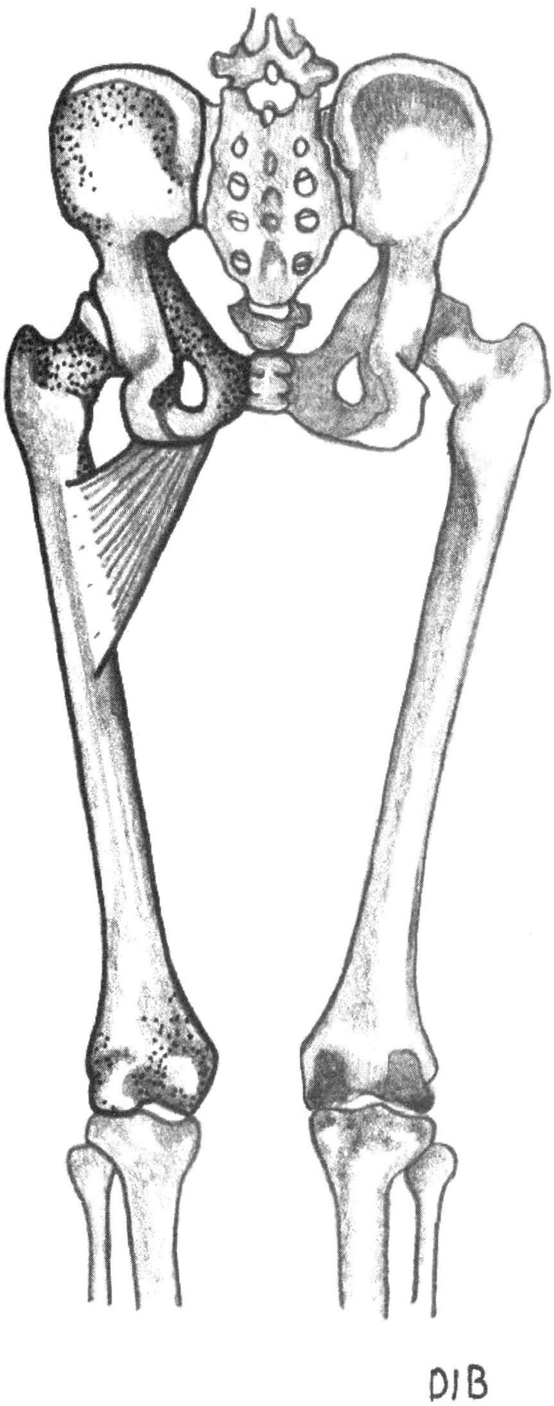

Origin:
Pubis (inferior ramus)

Insertion:
Femur (proximal 1/3 of medial lip of linea aspera and distal pectineal line)

Action:
Adduction of femur at hip,
Flexion of femur at hip (accessory),
Internal rotation of femur (accessory)

Innervation:
Obturator nerve (via spinal nerves L-2 thru L-4)

Relationship:
Adductor brevis lies under the pectineus and adductor longus with its fibers coursing laterally and posteriorly as it broadens and descends toward its distal attachment. It is intermediately located posterior to longus and anterior to magnus. Adductor brevis is synergistic in adduction with longus, magnus, pectineus & gracilis, and is opposed in that action by gluteus medius & minimus, and tensor fascia latae. In internal (medial) rotation, the adductor group acts with the anterior portion of gluteus minimus and is opposed by the deep lateral rotators and gluteus maximus. Adductor brevis is active in walking, running, jumping, and stair climbing. It is strengthened by adduction against resistance. Symptoms commonly include knee, thigh & groin pain (worse with weight bearing & twisting of hip), and decreased ROM in abduction & lateral rotation. Pain is increased by strenuous activity and overuse. Trigger point pain may be referred upward to deep in the groin and downward to the knee and shin. Symptoms must be differentiated from those of vastus medialis & pectineus dysfunction, lumbar disc disease, and nerve entrapment.

ADDUCTOR MAGNUS (right)
Anterior View

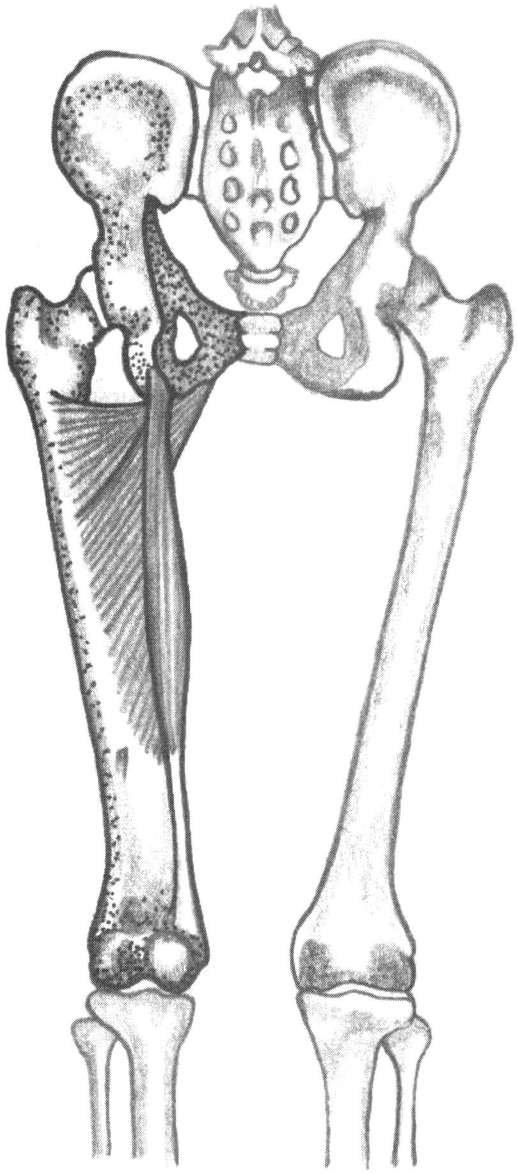

Origin:
Pubis (inferior ramus),
Ischium (inferior ramus and inferior and lateral aspect of ischial tuberosity)

Insertion:
Femur (whole length of linea aspera by an aponeurosis, medial supracondylar line, adductor tubercle on medial condyle)

Action:
Adduction of femur at hip,
Flexion of femur at hip (accessory),
Internal rotation of femur (accessory),
Extension of femur at hip (posterior fibers)

Innervation:
Obturator nerve (superior and middle fibers),
& sciatic nerve (via spinal nerves L-1 thru S-1)

Relationship:
Adductor magnus is the largest and most posterior of the adductor group. This muscle is located on the medial thigh and appears as three distinct bundles. The superior fibers from the pubic ramus are short and horizontal. Occasionally, these fibers are inserted into a line from the greater trochanter to the linea aspera and seem to form a distinct separate muscle. When this occurs, the muscle is called the adductor minimus. The medial fibers move downward and lateral. The most distal bundle descends almost vertically to a tendon on the distal 1/3 of the thigh. Adductor magnus is active in walking, running, jumping, skiing, and horse riding. It is strengthened by adduction against resistance. Symptoms commonly include deep intrapelvic pain, anterior thigh and groin pain (worse lying on involved side). Trigger point pain may be referred from the groin, down the anteriomedial thigh to the knee. Symptoms must be differentiated from those of vastus medialis & pectineus dysfunction, adductor avulsion, lumbar disc disease, pubic fracture or inflammation, prostatitis, hernia, and nerve entrapment.

GRACLILIS (left)
Anterior View

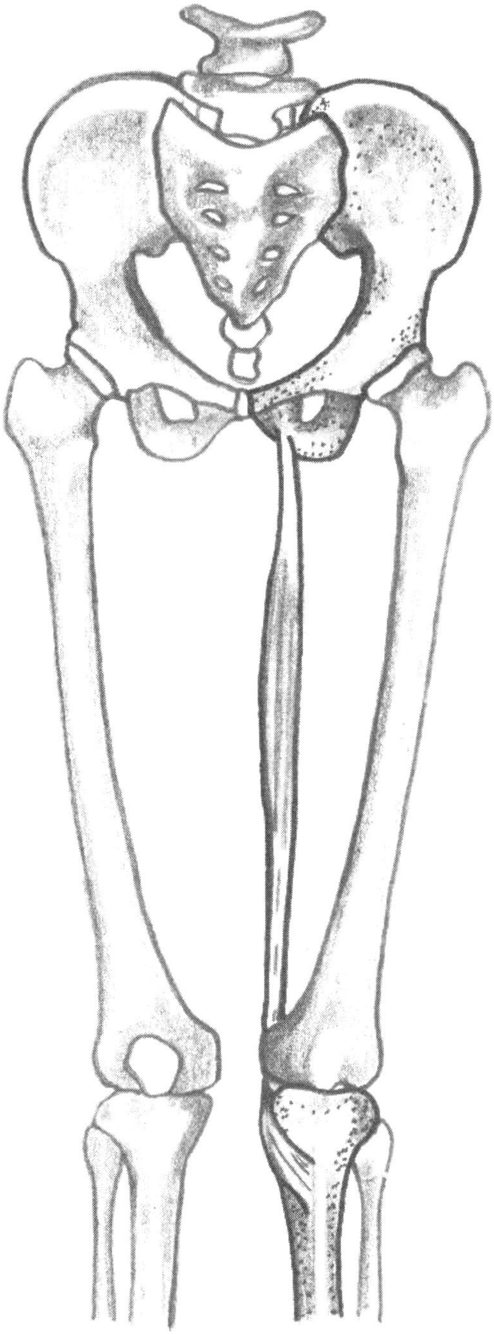

Origin:
Pubis (inferior ramus near symphysis)

Insertion:
Tibia (medial surface of shaft distal to the tibial condyle)

Action:
Adduction of femur at hip
Flexion of femur at hip (accessory)
Flexion of knee (from extension)
Internal rotation of tibia (with knee flexed)

Innervation:
Obturator nerve (anterior division, via spinal nerves L-2 & L-3)

Relationship:
Gracilis is the second longest muscle in the body. It lies most superficially on the medial thigh as a thin, flat muscle that tapers and narrows distally. The fibers are directed vertically and join a tendon that curves around the medial condyle of the femur and then around the medial condyle of the tibia. Its tendon is one of three (along with those of the sartorius and semitendinosus) that unite to form the pes anserinus. In adduction, gracilis acts synergistically with the three adductors and pectineus. In flexion of the knee and medial rotation, it acts with the hamstrings. Gracilis is active in walking, running, jumping, batting, horseback riding, and serving (in tennis). It is strengthened by adduction against resistance. The chief symptom of gracilis is superficial, hot, stinging pain in the medial thigh. It may also be described as a diffuse, localized aching. The pain of gracilis is unchanged by rest, but is ameliorated by walking or by not changing position. Trigger point pain is superficial, and noted along the anterior thigh. Symptoms must be differentiated from those of vastus medialis & pectineus dysfunction, lumbar disc disease, nerve entrapment, hernia, prostatitis, and symphysitis.

TIBIALIS ANTERIOR (left)
Anterolateral View

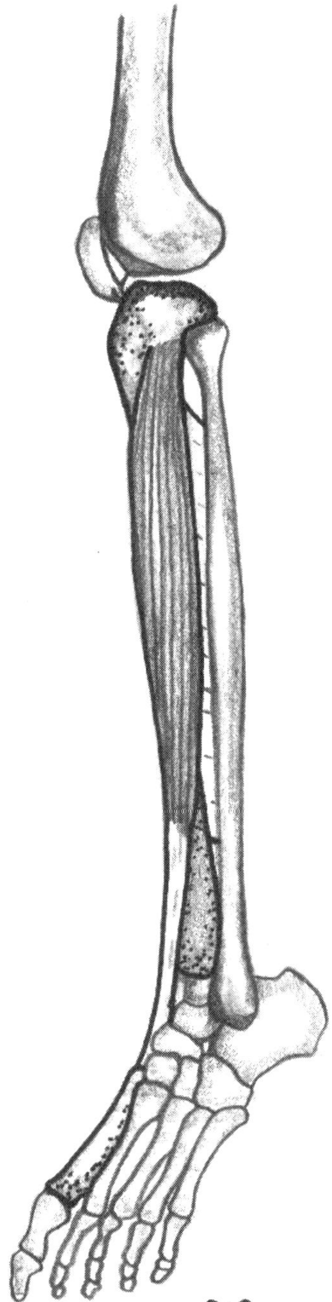

Origin:
Tibia (lateral condyle and proximal 2/3 of lateral surface),
Interosseous membrane,
Deep surface of crural fascia

Insertion:
1st (medial) cuneiform (on medial and plantar surfaces),
1st metatarsal (base)

Action:
Ankle dorsiflexion (at talocrural joint),
Foot inversion and adduction (supination) at subtalar and midtarsal joints

Innervation:
Deep peroneal nerve (via spinal nerves L-4 thru S-1)

Relationship:
Tibialis anterior is located on the lateral aspect of the tibia. It has a thick belly proximally, becoming thin and tendinous distally. The fibers pass vertically downward and end in a prominent tendon on the anterior surface of the lower leg. It is contained in the most medial compartments of the extensor retinaculum. Tibialis anterior functions to maintain balance during standing and is very active during walking, running, and jumping. It is only slightly to moderately used in cycling and throwing. Tibialis anterior may be strengthened by dorsiflexion and inversion against resistance, as well as by walking barefoot (inverted) in soft sand. The major complaints associated with tibialis anterior are local pain, weakness of dorsiflexion when walking and ankle pain without joint injury. Trigger points refer pain to the anteriomedial aspect of the ankle and down to the great toe. Symptoms must be differentiated from those of L-5 radiculopathy, anterior compartment syndrome (see page 152), and those of other muscles including the flexors and peroneus tertius.

EXTENSOR DIGITORUM LONGUS (left)
Anterolateral View

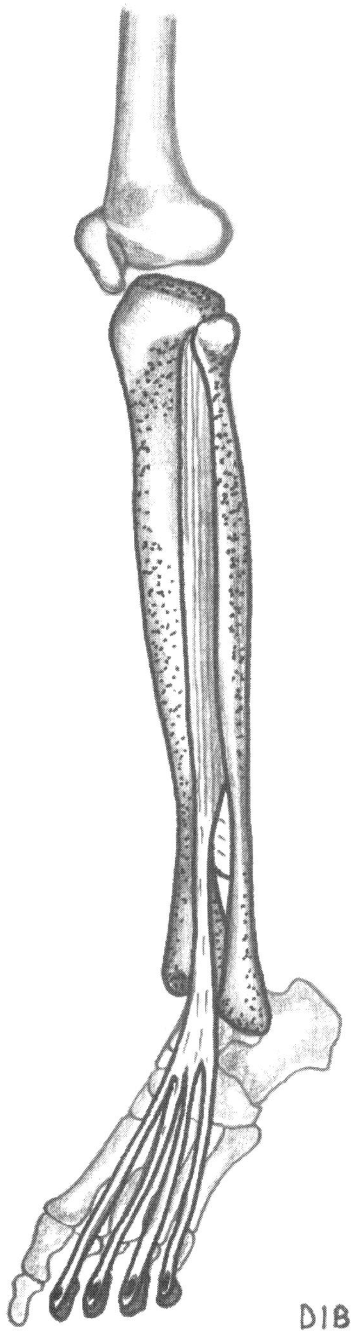

Origin:
Tibia (lateral condyle on lateral side),
Fibula (shaft, upper 3/4 of anterior surface),
Interosseous membrane (proximal portion),
Deep fascia and intermuscular septum

Insertion:
Tendon of insertion divides into four tendons to dorsum of foot that form an expansion over each toe,
Toes 2 thru 5: Proximal phalanges of the four lesser toes (intermediate slip to dorsum of base of each),
Middle and distal phalanges (two lateral slips to dorsum of base of each)

Action:
Extension of four lesser toes, proximal phalanx (forcefully), and middle & distal phalanx (weakly),
Ankle dorsiflexion (accessory),
Foot eversion (accessory)

Innervation:
Deep peroneal nerve (via spinal nerves L-4 thru S-1)

Relationship:
This muscle lies in the lateral aspect of the anterior leg. It descends lateral to the tibialis anterior, and its distal tendon accompanies the tendon of the peroneus tertius before dividing. Extensor digitorum longus is active during walking, running and jumping. It is also active on the contralateral side when throwing (baseball) or swinging (tennis, golf, etc.). The major antagonists are gastrocnemius, soleus, and the peroneals. This muscle may be strengthened by dorsiflexion and toe extension against resistance. Major complaints include pain in the dorsum of the foot, weakness when walking, and nocturnal cramping of the toes. This muscle may also be involved in hammer toe and claw toe deformities. Trigger points refer pain to the dorsolateral aspect of the foot and down to the three middle toes. Symptoms must be differentiated from those arising in the tarsal joints or other nearby muscles.

EXTENSOR HALLUCIS LONGUS (left)

Anterolateral View

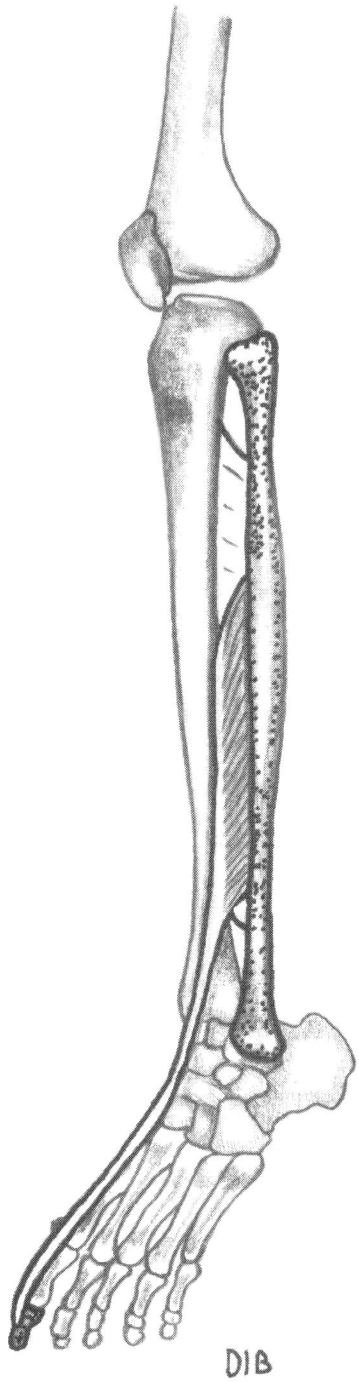

Origin:
Fibula (medial surface, along middle 1/2 of body),
Interosseous membrane

Insertion:
Hallux (base of distal phalanx),
Expansion to base of proximal phalanx

Action:
Hallux extension (proximal phalanx powerfully),
Ankle dorsiflexion (accessory),
Foot inversion (accessory)

Innervation:
Deep peroneal nerve (via spinal nerves L-4 thru S-1)

Relationship:
This thin muscle descends in the leg between and largely covered by the tibialis anterior and the extensor digitorum longus. Its tendon does not emerge superficially until it reaches the distal 1/3 of the leg. Like extensor digitorum longus, this muscle is active during walking, running and jumping. It is also active on the contralateral side when throwing (baseball) or swinging (tennis, golf, etc.). This muscle may be strengthened by dorsiflexion, inversion, and hallux extension against resistance. Associated complaints include localized foot pain, nocturnal cramping, weakness when walking, or during dorsiflexion. Trigger points refer pain to the dorsomedial aspect of the foot and down to the great toe. Symptoms must be differentiated from those arising in the tarsal joints or other nearby muscles.

PERONEUS LONGUS (left)
Anterolateral View

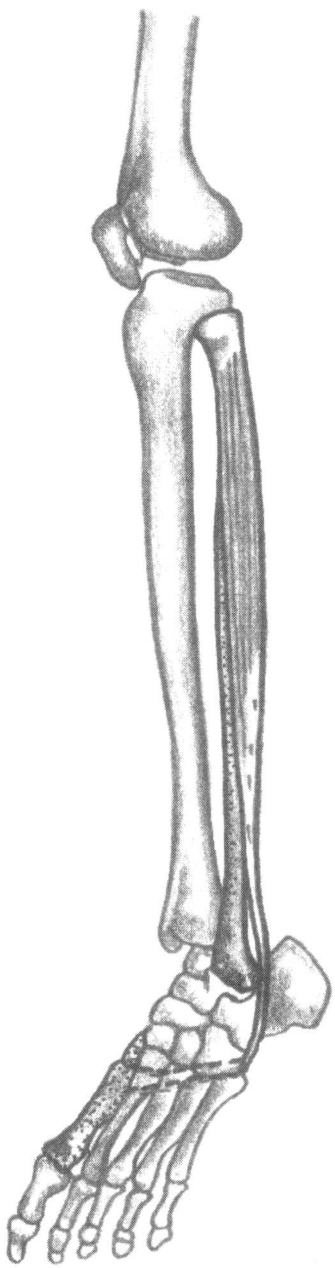

Origin:
Fibula (head and upper 2/3 of lateral shaft),
Tibia (lateral condyle, occasionally),
Deep fascia and intermuscular septa

Insertion:
1st metatarsal (lateral plantar side of base),
1st (medial) cuneiform (lateral plantar aspect),
2nd metatarsal (occasionally by a slip)

Action:
Foot eversion (with abduction, thus pronation),
Ankle plantar flexion (assist),
Depression of 1st metatarsal,
Support of transverse arch

Innervation:
Superficial peroneal nerve (via spinal nerves L-4 thru S-1)

Relationship:
This muscle is found proximally on the fibular side of the leg where it is superficial to peroneus brevis. The belly ends in a long tendon that passes behind the lateral malleolus (with the brevis) and runs obliquely forward to the base of both the 1st metatarsal and the medial cuneiform bones. Peroneus longus functions eccentrically to control movement in deceleration. It is active in standing, walking, and running. Peroneus longus is opposed (in eversion) by tibialis anterior and posterior. This muscle is best strengthened by walking barefoot (everted) in soft sand. Complaints associated with peroneus longus include local pain, weak ankles (which contribute to instability and frequent eversion sprains), and foot pain with callusing. Flat feet and Morton's Foot (shorter first and longer second metatarsal bone) may contribute to peroneus dysfunction. Trigger points refer pain to the area of the lateral malleolus and down the lateral aspect of the foot. Symptoms must be differentiated from those of the tarsal joints, ankle sprain, lateral compartment syndrome, referred pain from lower limb extensor muscles, and entrapment of the peroneal nerves.

PERONEUS BREVIS (right)
Anterolateral View

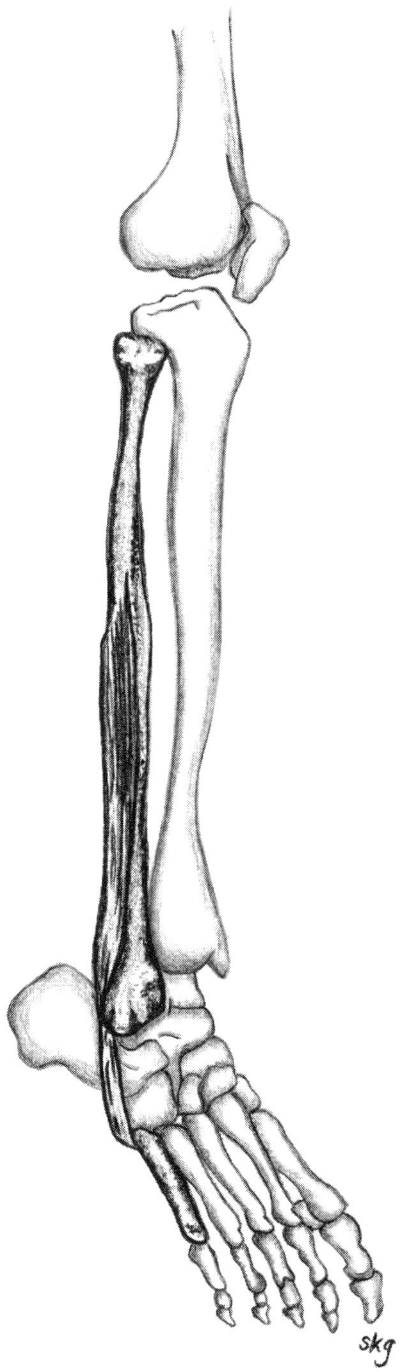

Origin:
Fibula (distal 2/3 of lateral surface of shaft), Intermuscular septa

Insertion:
5th metatarsal (tuberosity on lateral surface of base)

Action:
Foot eversion (with abduction, thus pronation), Ankle plantar flexion (accessory)

Innervation:
Superficial peroneal nerve (via spinal nerves L-4 thru S-1)

Relationship:
The Peroneus brevis lies deep to the longus and is the shorter and smaller muscle of the two. The belly fibers descend vertically to end in a tendon, which courses (with the longus) behind the lateral malleolus to its insertion. The pair of muscles share a synovial sheath. Peroneus brevis exerts only a small percentage of the effort of longus. It is (somewhat) active in standing, walking, and running. It is likewise opposed (in eversion) by tibialis anterior and posterior. This muscle is best strengthened by eversion against resistance. Complaints associated with peroneus brevis include local pain, weak ankles (which contribute to instability & frequent eversion sprains), and foot pain with callusing. Flat feet and Morton's Foot (shorter first and longer second metatarsal bone) may contribute to peroneus dysfunction. Trigger points refer pain to the area of the lateral malleolus and down the lateral aspect of the foot. Symptoms must be differentiated from those of the tarsal joints, ankle sprain, lateral compartment syndrome, referred pain from lower limb extensor muscles, and entrapment of the peroneal nerves.

PERONEUS TERTIUS (left)
Anterolateral View

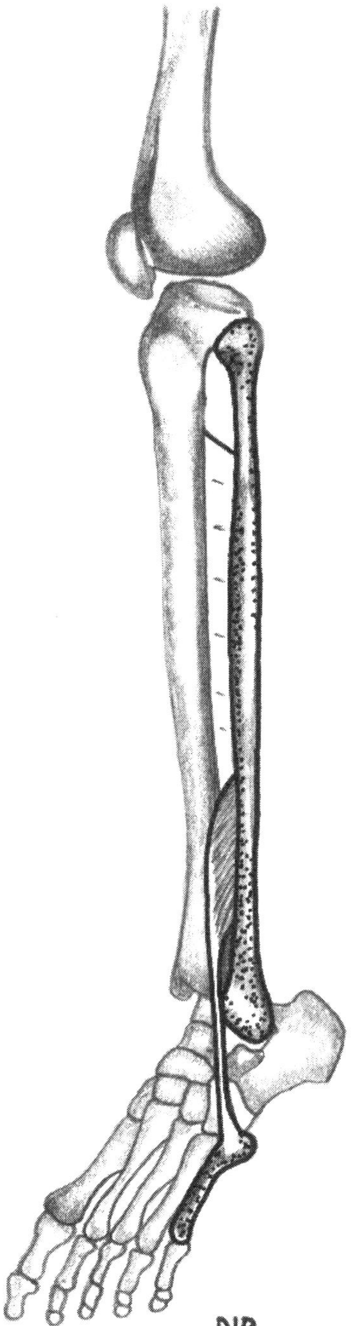

Origin:
Fibula (distal 1/3 to 1/2 of medial surface),
Interosseous membrane,
Intermuscular septum

Insertion:
5th metatarsal (tubercle at dorsal surface of base)

Action:
Ankle dorsiflexion,
Foot eversion (accessory)

Innervation:
Deep peroneal nerve (via spinal nerves L-5 & S-1)

Relationship:
This muscle is considered to be the inferior and lateral portion of extensor digitorum longus (the fifth tendon). The muscle descends on the lateral leg, diving under the extensor retinaculum in the same passage as extensor digitorum longus, to insert on the 5th metatarsal. It is distinct from the other peroneus muscles in that its tendon passes anterior to the lateral malleolus and thus acts in opposition to longus and brevis during plantar flexion. It is active in standing, walking, and running. This muscle is best strengthened by eversion against resistance. Complaints associated with peroneus tertius include local pain (worse with active dorsiflexion), weak ankles (which contribute to instability and frequent eversion sprains), and foot pain with callusing. Trigger points refer pain to the anterolateral aspect of the ankle. Symptoms must be differentiated from those of the tarsal joints, ankle sprain, lateral compartment syndrome, referred pain from lower limb extensor muscles, and entrapment of the peroneal nerves.

EXTENSOR DIGITORUM BREVIS (left)
W/ EXTENSOR HALLUCIS BREVIS
Anterolateral View

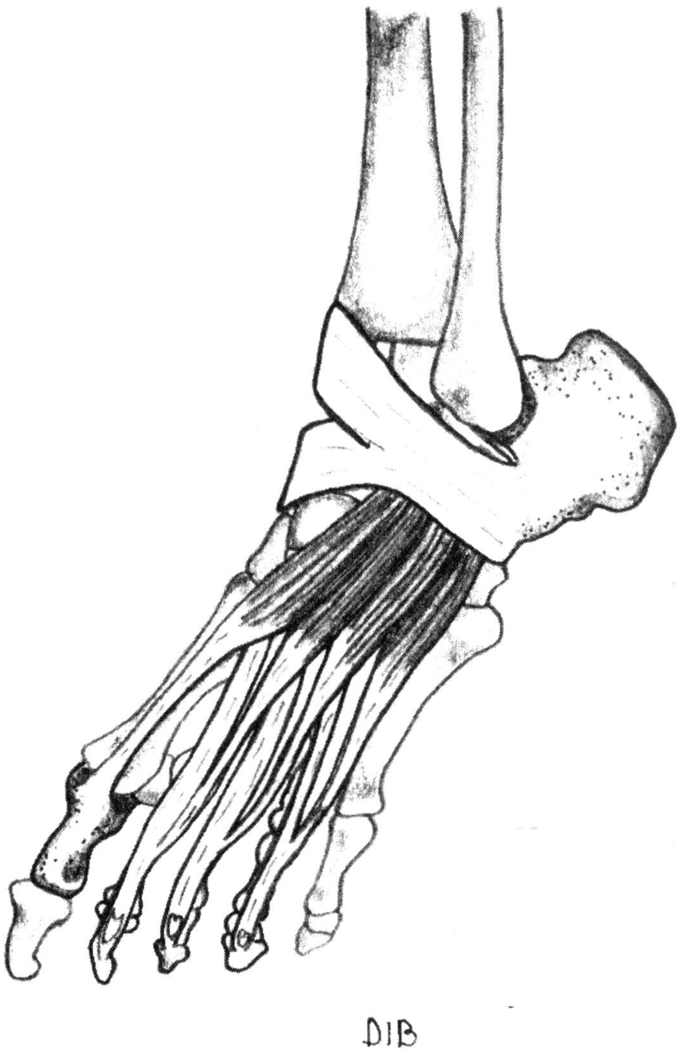

Origin:
Anterior & lateral surfaces of the Calcaneus, lateral talocalcaneal ligament, inferior extensor retinaculum

Insertion:
Base of proximal phalanx of great toe, into the sides of the extensor digitorum longus tendons of the 2nd, 3rd, and 4th phalanges

Action:
Extension of proximal phalanx of great toe (via extensor hallucis brevis) and all three phalanges of the 2nd, 3 rd, and 4th toes

Innervation:
Deep peroneal nerve (via spinal nerves L-5 & S-1)

Relationship:
Extensor digitorum (and hallucis) brevis are active during walking, running and jumping. They serve during ambulation, to provide both rigidity for propulsion and flexibility for shock absorption. This muscle is best strengthened by extension of the toes against resistance. Complaints associated with extensor digitorum (and hallucis) brevis often include deep aching pain in the dorsum of the foot, weakness when walking, and subjects may tend to have an antalgic gait. Active trigger points may present as foot cramps or pain referred to the instep. Symptoms must be differentiated from those of flat feet, bunions, avulsion & stress fractures, and arthritis.

INTRINSIC MUSCLES of the FOOT (right)
Plantar View

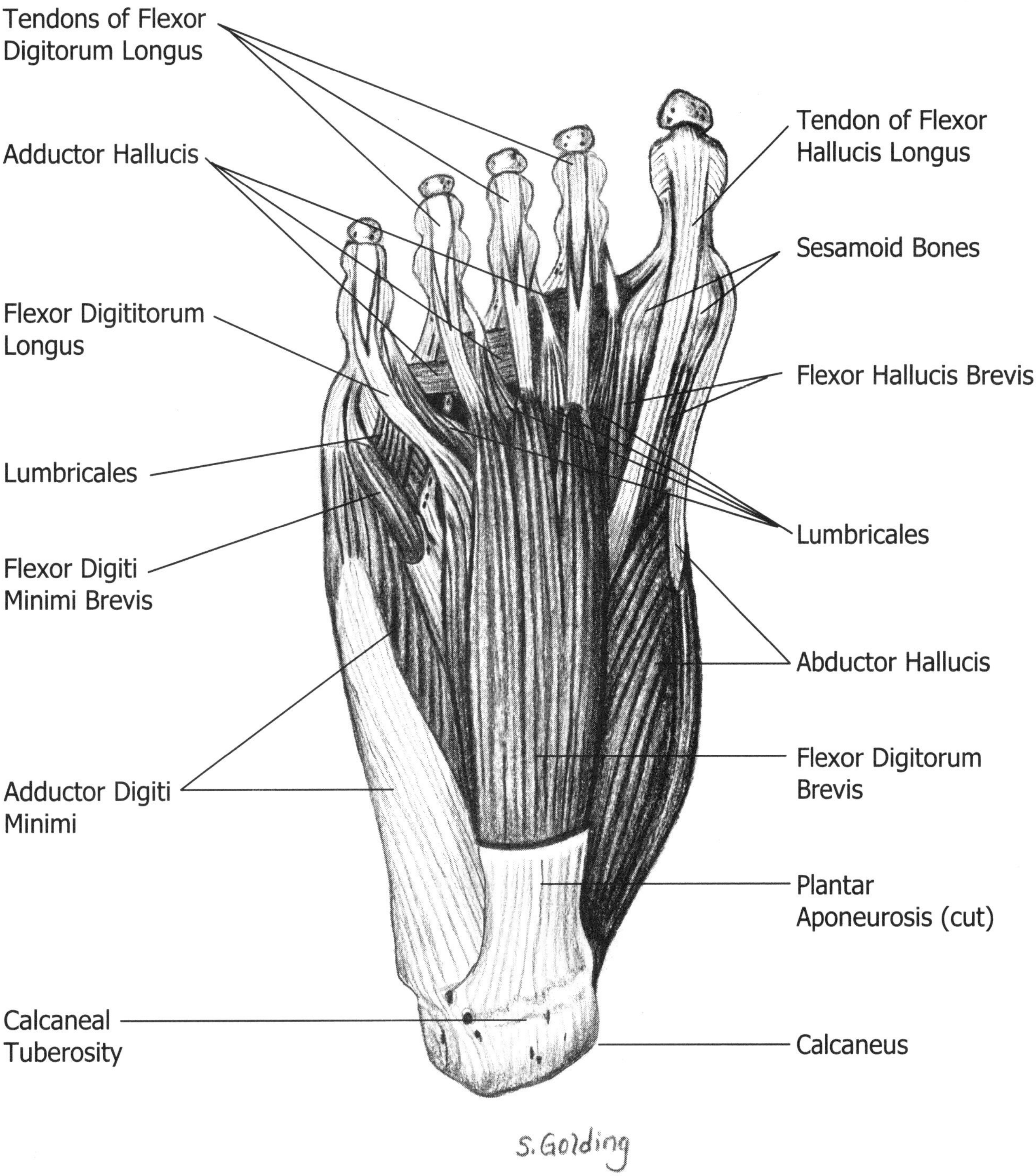

Transverse Section of (left) Leg
Superior View (at mid leg)

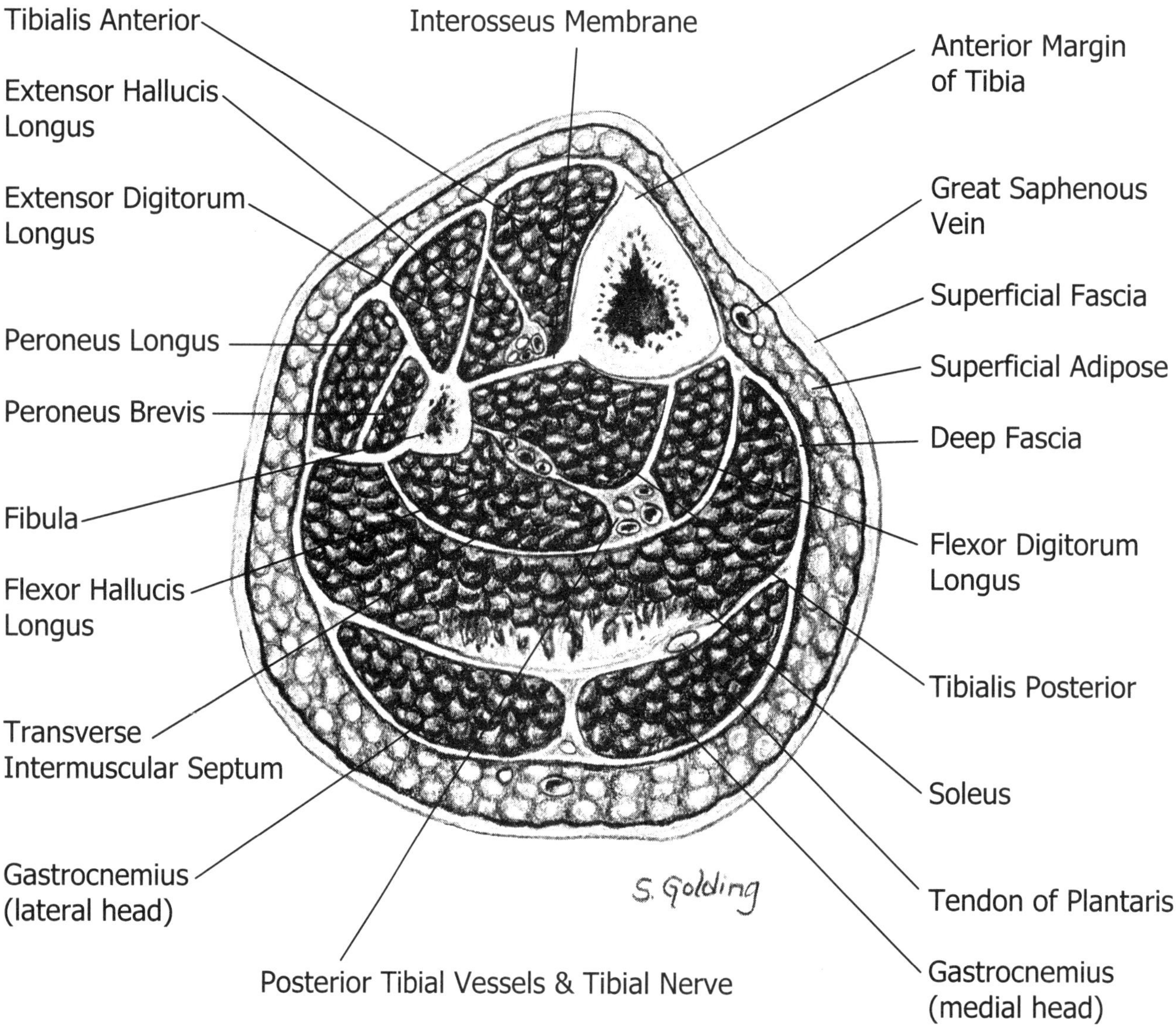

The major structures of the lower leg include the: tibia, fibula, an anterior (extensor) group of muscles, a posterior (flexor) group of muscles, and a lateral (evertor - peronei) group of muscles. These muscle groups are separated and encapsulated by deep fascial septae (the fascia cruris) that is essentially a continuation of the fascia lata. The largest of these three compartments is the posterior, which contains the bulky and powerful plantar flexors.

SUBOCCIPITALS

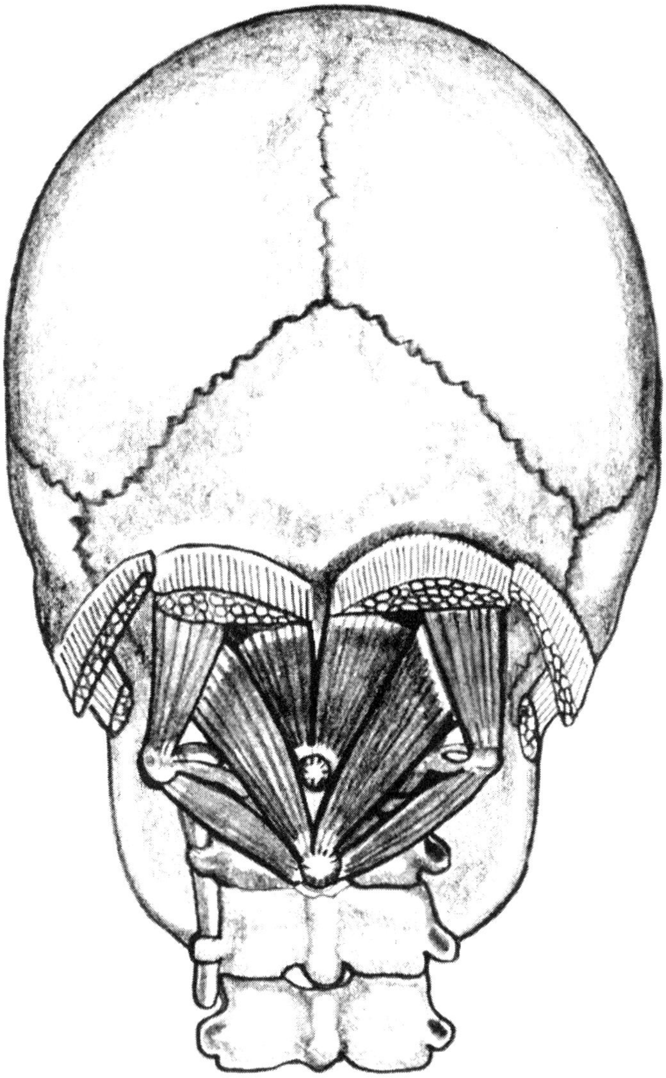

DIB

The four-paired **Suboccipital** muscles are the deepest muscles found just inferior to the base of the skull (occipital bone.) Three of the four muscles are attached to the occiput (the fourth being attached superiorly to the transverse processes of C-1). The shared function of these muscles is to provide and control nodding, rotation, and sidebending movements of the head. Direct palpation is difficult, yet dysfunction is displayed and identified by specific restrictions to capital flexion, rotation, and sidebending.

RECTUS CAPITIS POSTERIOR MAJOR
Description: Originating as a small tendon, it broadens as it rises superiorly and laterally.

RECTUS CAPITUS POSTERIOR MINOR
Description: Originating as a narrow tendon, broadening into a wide band, this short muscle ascends nearly vertically to its attachment.

OBLIQUUS CAPITIS SUPERIOR
Description: Originating as a narrow muscle, it widens as it rises superiorly, medially, and nearly vertically to its attachment.

OBLIQUUS CAPITIS INFERIOR
Description: Originating medially, it passes laterally and slightly superior to its attachment. It is the larger of two obliqui and the only suboccipital not to have an attachment on the skull.

RECTUS CAPITIS POSTERIOR MAJOR

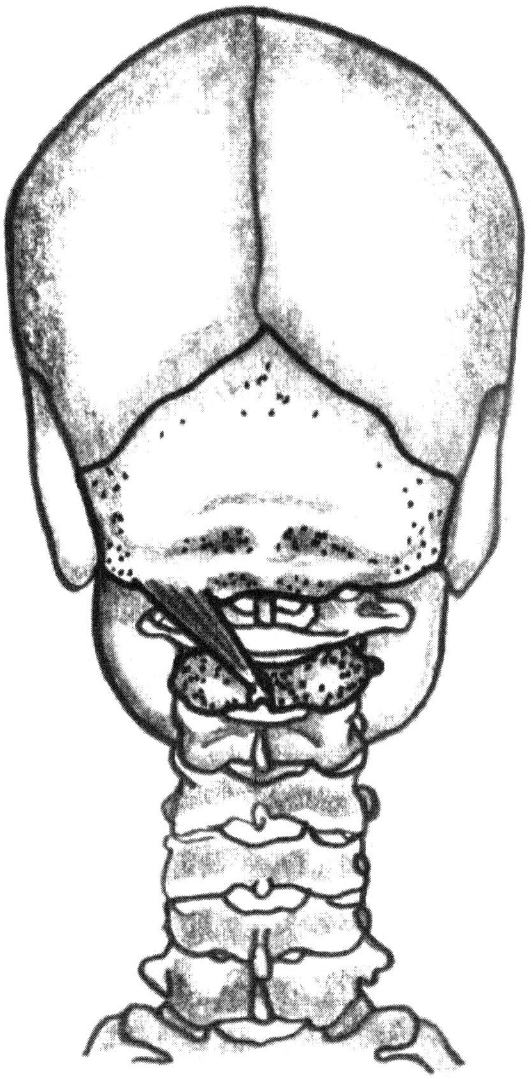

Origin:
Spinous process of vertebra C-2 (Axis)

Insertion:
Occiput (lateral part of inferior nuchal line and surface of bone just inferior to nuchal line)

Action:
Capital (head) extension,
Rotation of the head to the same side,
Lateral bending of the head to same side (minimal)

Innervation:
Suboccipital nerve (via spinal nerve C-1, branches of dorsal rami)

Relationship:
RCPM originates as a small tendon that broadens as it rises superiorly and laterally to insert. This muscle is synergistic with semispinalis capitus in capital extension, with splenius capitus (same side) and SCM (opposite side) in rotation, and with rectus capitus lateralis in sidebending. It is opposed by longus capitus and rectus capitus anterior in capital extension and by its opposite side counterparts in rotation and sidebending. This muscle is also very active in checkreining forward motion of the head. Symptoms related to rectus capitus posterior major include local pain, headache (made worse by lying down, that is described as being deep in the upper neck), and a noticeably restricted range of motion. Complaints are aggravated by chronic shortening (head forward posture or maintaining an upward/ sideways gaze), cold temperature, and trauma. Trigger point pain is referred from the occiput toward the orbit. Symptoms must be differentiated from those of chronic intractable benign pain, tension headache, cervicogenic headache, articular dysfunction, and occipital neuralgia.

RECTUS CAPITIS POSTERIOR MINOR

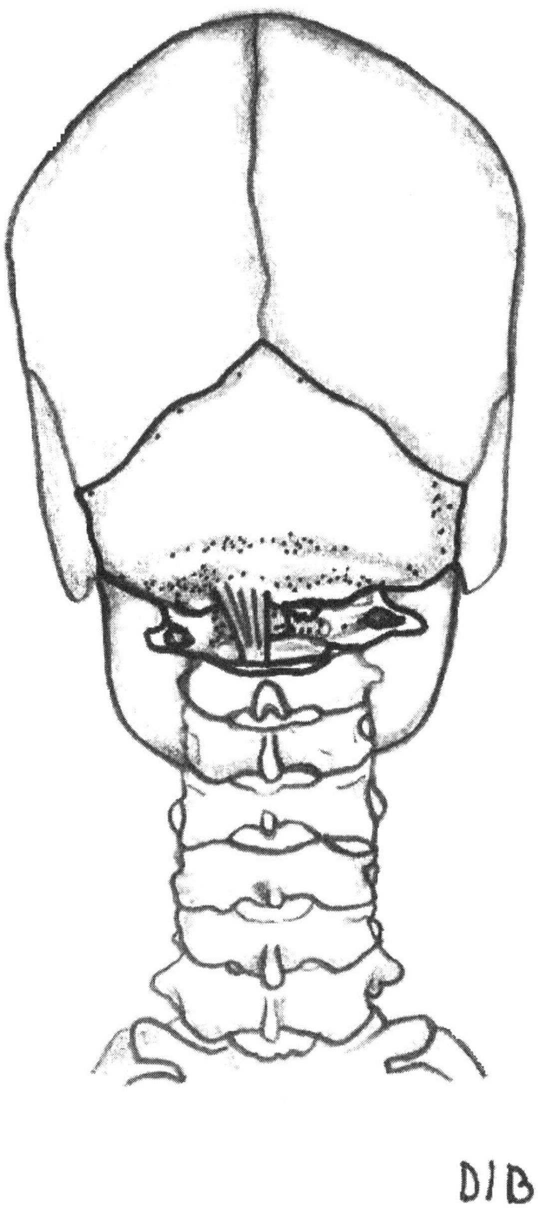

Origin:
Tubercle on posterior arch of vertebra C-1 (Atlas)

Insertion:
Occiput (medial portion of inferior nuchal line and surface between the inferior nuchal line and the foramen magnum)

Action:
Capital (head) extension,
Rotation of the head to the same side,
Lateral bending of the head to same side (minimal)

Innervation: Suboccipital nerve (via spinal nerve C-1, branches of dorsal rami)

Relationship:
RCPm originates as a narrow tendon, broadening into a wide band as it ascends to its attachment. It is synergistic with semispinalis capitus in capital extension, with splenius capitus (same side) & SCM (opposite side) in rotation, and with rectus capitus lateralis in sidebending. It is opposed by longus capitus and rectus capitus anterior in capital extension and by its opposite side counterparts in rotation and sidebending. This muscle is also very active in checkreining forward motion of the head. Symptoms related to rectus capitus posterior minor include local pain, headache (made worse by lying down, that is described as being deep in the upper neck), and a noticeably restricted range of motion. Complaints are aggravated by chronic shortening (head forward posture or maintaining an upward/ sideways gaze,) cold temperature, and trauma. Trigger point pain is referred from the occiput toward the orbit. Symptoms must be differentiated from those of chronic intractable benign pain, tension headache, cervicogenic headache, articular dysfunction, and occipital neuralgia.

OBLIQUUS CAPITIS SUPERIOR

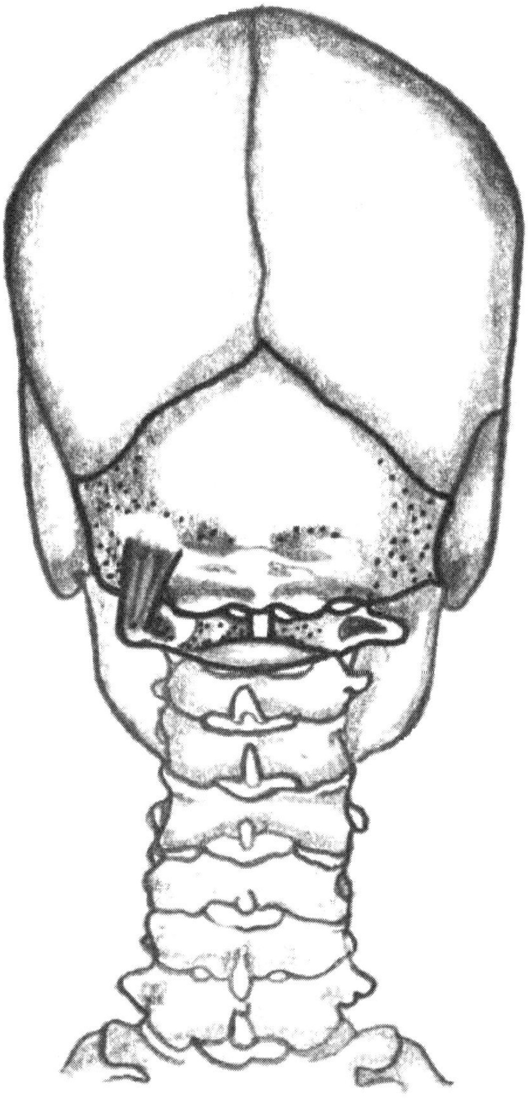

Origin:
Atlas, C-1 (superior surface of the transverse process) sharing insertion with obliquus capitis inferior

Insertion:
Occiput (between superior and inferior nuchal lines & deep to the lateral portion of semispinalis capitis).

Action:
Capital (head) extension,
Rotation of the head to the same side,
Lateral bending of the head to same side (minimal)

Innervation:
Suboccipital nerve (via spinal nerve C-1, branches of dorsal rami)

Relationship:
OCS starts as a narrow muscle, widening as it rises superiorly and (slightly) medially toward its attachment. It is synergistic with semispinalis capitus in capital extension, with splenius capitus (same side) and SCM (opposite side) in rotation, and with rectus capitus lateralis in sidebending. It is opposed by longus capitus and rectus capitus anterior in capital extension and by its opposite side counterparts in rotation and sidebending. This muscle is also very active in checkreining forward motion of the head. Symptoms related to obliquus capitus superior include local pain, headache (made worse by lying down, that is described as being deep in the upper neck), and a noticeably restricted range of motion. Complaints are aggravated by chronic shortening (head forward posture or maintaining an upward/sideways gaze), cold temperature, and trauma. Trigger point pain is referred from the occiput toward the orbit. Symptoms must be differentiated from those of chronic intractable benign pain, tension headache, cervicogenic headache, articular dysfunction, and occipital neuralgia.

OBLIQUUS CAPITIS INFERIOR

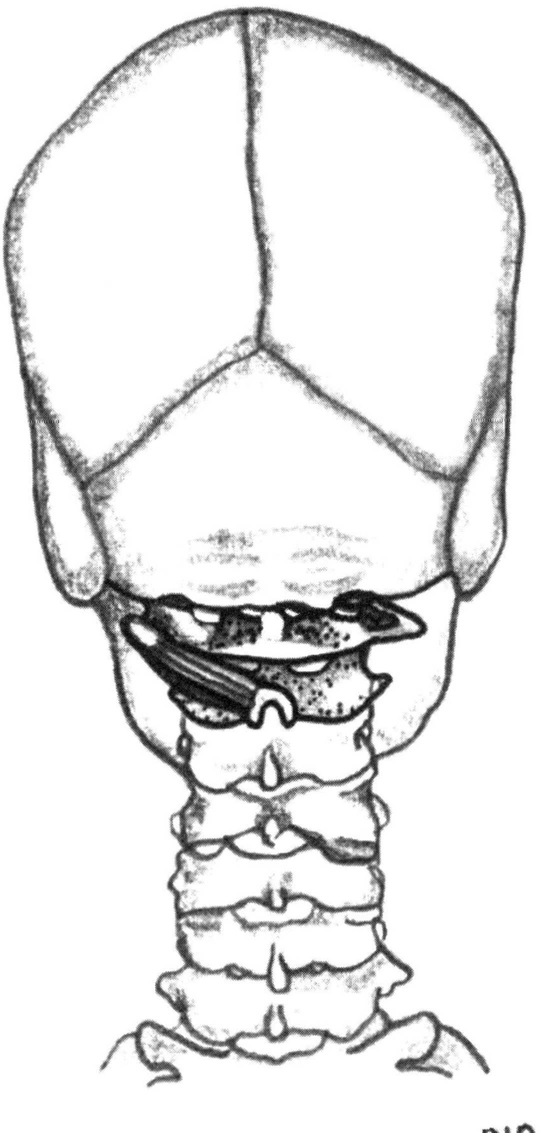

Origin:
Axis, C-2 (apex of spinous process)

Insertion:
Atlas, C-1 (inferior and dorsal aspect of transverse process)

Action:
Capital (head) extension,
Rotation of the head to the same side,
Lateral bending of the head to same side (minimal)

Innervation:
Suboccipital nerve (via spinal nerve C-1, branches of the dorsal rami)

Relationship:
OCI passes laterally and slightly superior toward its attachment. It is the larger of the two obliqui.
This muscle is synergistic with semispinalis capitus in capital extension, with splenius capitus (same side) and SCM (opposite side) in rotation, and with rectus capitus lateralis in sidebending. It is opposed by longus capitus and rectus capitus anterior in capital extension and by its opposite side counterparts in rotation and sidebending. This muscle is also very active in checkreining forward motion of the head. Symptoms related to obliquus capitus inferior include local pain, headache (made worse by lying down) that is described as being deep in the upper neck and a noticeably restricted range of motion. Complaints are aggravated by chronic shortening (head forward posture or maintaining an upward/ sideways gaze,) cold temperature, and trauma. Trigger point pain is referred from the occiput toward the orbit. Symptoms must be differentiated from those of chronic intractable benign pain, tension headache, cervicogenic headache, articular dysfunction, and occipital neuralgia.

CRANIOFACIAL MUSCLES
Anterior View

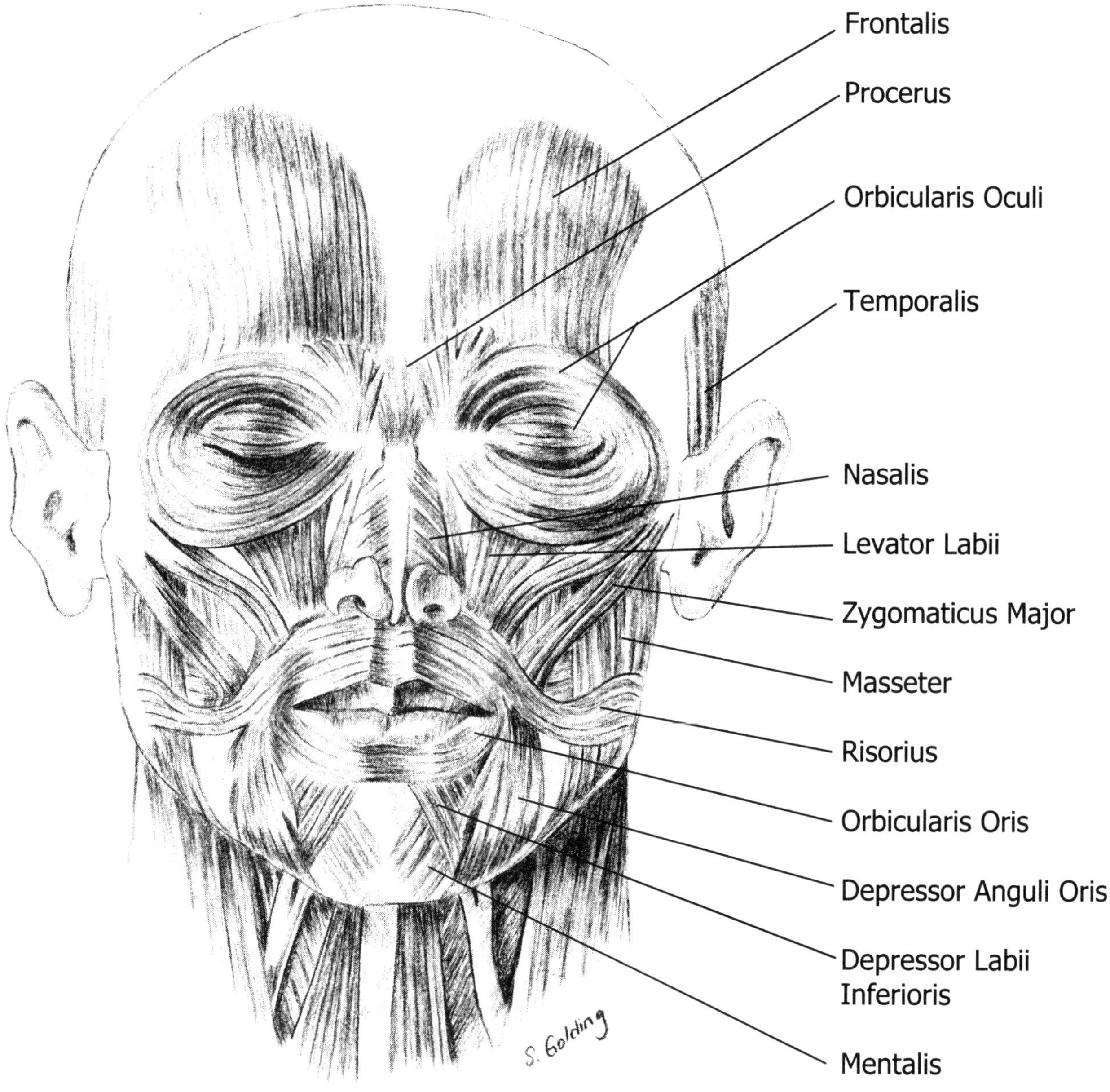

* See page 160 for Origin, Insertion, Action, and Innervation

CRANIOFACIAL MUSCLES
Lateral View

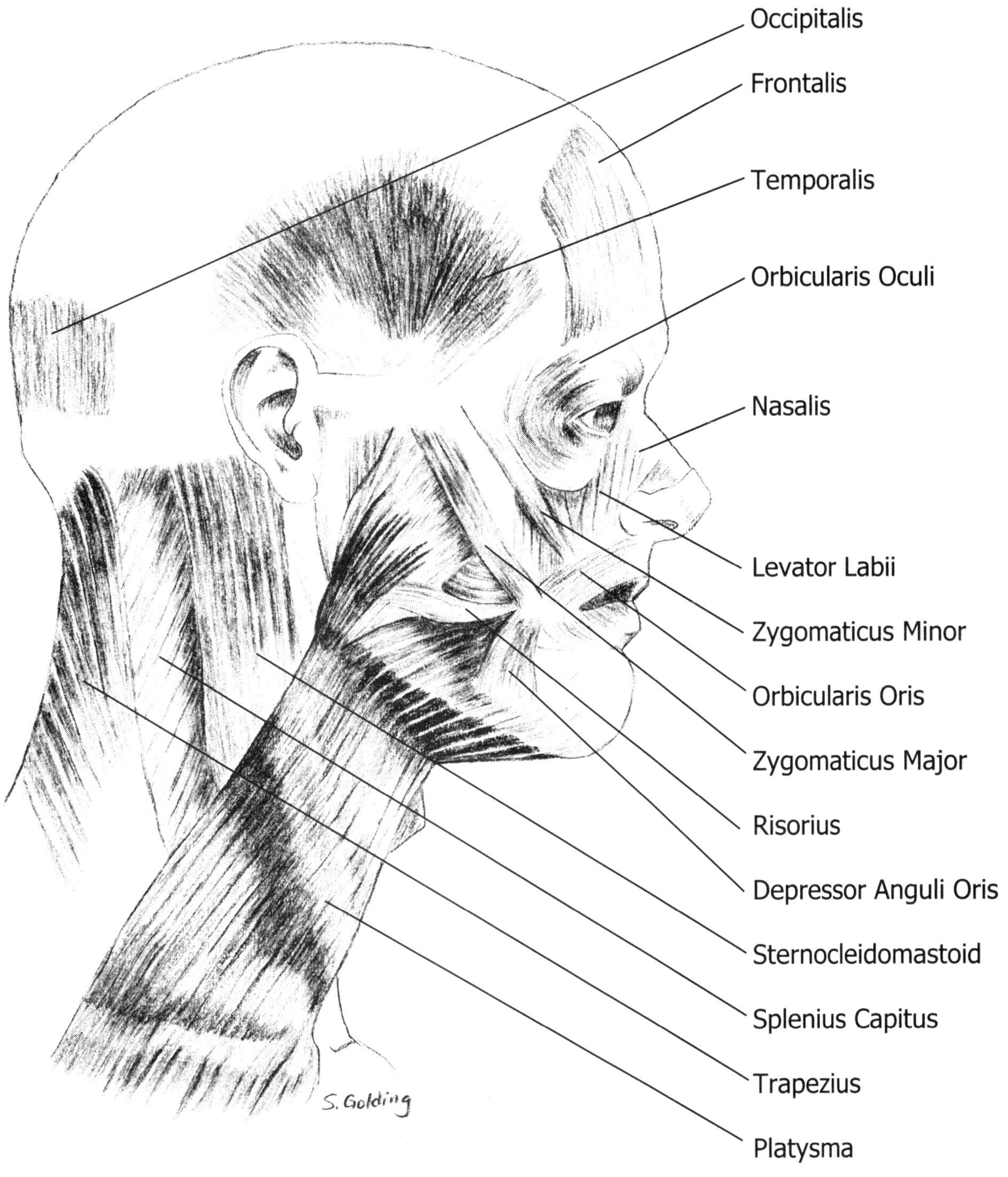

* See page 160 for Origin, Insertion, Action, and Innervation

CRANIOFACIAL MUSCLES

Occipitalis
Origin: Nuchal line of occiput & mastoid
Insertion: Epicranial aponeurosis
Action: Moves scalp posteriorly
Innervation: Cranial Nerve VII

Frontalis
Origin: Epicranial aponeurosis
Insertion: Superficial fascia of eyebrows
Action: Elevates eyebrows, wrinkles brow
Innervation: Cranial Nerve VII

Procerus
Origin: Nasal bone & cartilage
Insertion: Skin between eyebrows
Action: Elevates skin of nose
Innervation: Cranial Nerve VII

Orbicularis Oculi
Origin: Medial margin of orbit
Insertion: Eyelid & around eye
Action: Closes eyelid (& squinting)
Innervation: Cranial Nerve VII

Nasalis
Origin: Maxilla
Insertion: Nasal cartilage
Action: Dilates anterior nasal opening
Innervation: Cranial Nerve VII

Levator Labii
Origin: Infraorbital area of maxilla
Insertion: Skin at angle of mouth
Action: Elevates & everts upper lip
Innervation: Cranial Nerve VII

Orbicularis Oris
Origin: Muscle fibers around mouth
Insertion: Angle of mouth
Action: Closes & protrudes lower lips
Innervation: Cranial Nerve VII

Zygomaticus Minor
Origin: Lateral zygomatic bone
Insertion: Muscles of upper lip
Action: Elevates upper lip
Innervation: Cranial Nerve VII

Zygomaticus Major
Origin: Zygomatic bone
Insertion: Angle of mouth
Action: Draws mouth laterally & superiorly
Innervation: Cranial Nerve VII

Risorius
Origin: Fascia over Parotid gland
Insertion: Skin at angle of mouth
Action: Draws angle of mouth laterally
Innervation: Cranial Nerve VII

Depressor Anguli Oris
Origin: Mandible
Insertion: Angle of mouth
Action: Draws mouth laterally & inferiorly
Innervation: Cranial Nerve VII

Platysma
Origin: Fascia over pectoralis & deltoid
Insertion: Mandible, muscles of lower face
Action: Depresses mandible & lower face
Innervation: Cranial Nerve VII

Depressor Labii Inferioris
Origin: Mandible
Insertion: Skin of lower lip
Action: Depresses corners of mouth
Innervation: Cranial Nerve VII

Mentalis
Origin: Mandible & incisor fossa
Insertion: Superficial tissue of chin
Action: Elevates & protrudes lower lip
Innervation: Cranial Nerve VII

EXTRINSIC MUSCLES OF THE EYE
(left) Lateral View

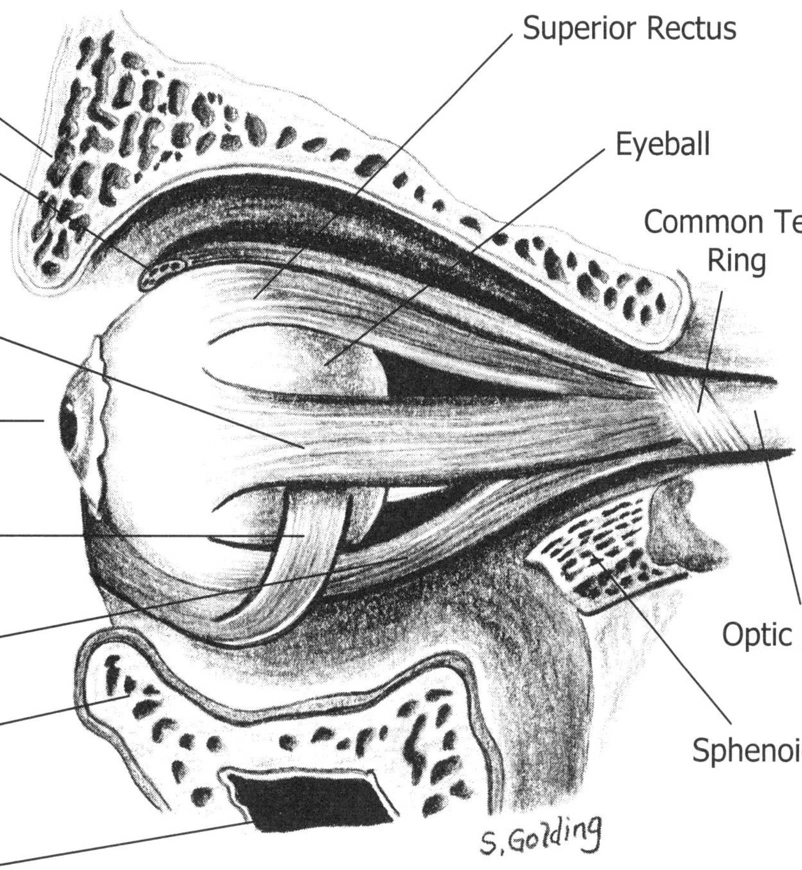

Superior Rectus
Origin:Common Tendinous Ring
Insertion: Superior/medial aspect of eyeball
Action: Rotates eyeball upward & medially
Innervation: Cranial Nerve III

Lateral Rectus
Origin:Common Tendinous Ring
Insertion: Lateral aspect of eyeball
Action: Rotates (abducts) eyeball laterally
Innervation: Cranial Nerve VI

Inferior Oblique
Origin: Maxilla at floor of orbit
Insertion: Between inferior & lateral recti
Action: Rotates eyeball upward & laterally
Innervation: Cranial Nerve III

Inferior Rectus
Origin: Common Tendinous Ring
Insertion: Inferior/medial aspect of eyeball
Action: Rotates eyeball downward & medially
Innervation: Cranial Nerve III

* **Superior Oblique** and **Medial Rectus** muscles are not illustrated.

BUCCINATOR (right)
Lateral View

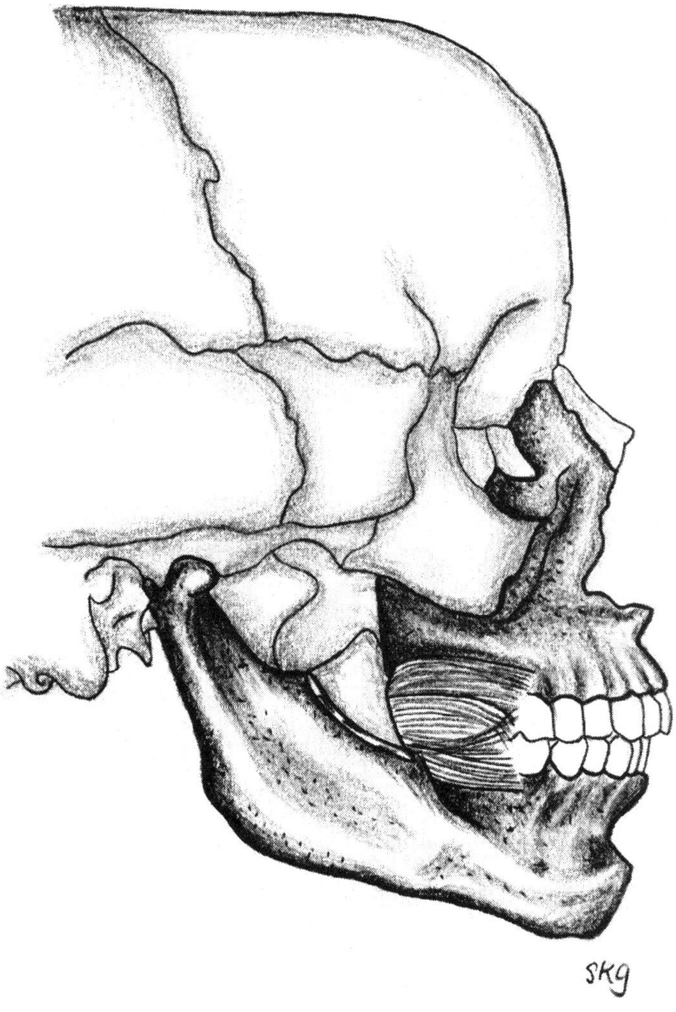

Origin:
Maxilla and mandible (external surfaces of the alveolar processes)

Insertion:
Deep portion of the orbicularis oris (muscles of the lips)

Action:
Compresses cheek against the teeth, Expels air when cheeks are distended (whistling & blowing a wind instrument), Acts in mastication to control passage of food (maintains food between teeth, in conjunction with the tongue)

Innervation:
Facial Nerve (via cranial nerve VII, buccal branch)

Description:
The principal muscle of the cheek is classified as a facial muscle (because of its innervation) despite its role in mastication. The buccinator forms the lateral wall of the oral cavity, lying deep to the other facial muscles, and filling the gap between the maxilla and the mandible. Complaints associated with buccinator include subzygomatic pain (worse with chewing) and a perception of difficulty in swallowing. Trigger point pain is referred to deep in the cheek as an ache in the jaw. Symptoms must be differentiated from those of tension headache, TMJ dysfunction, and SCM or scalene dysfunction.

PTERYGOID LATERALIS (right)
Lateral View

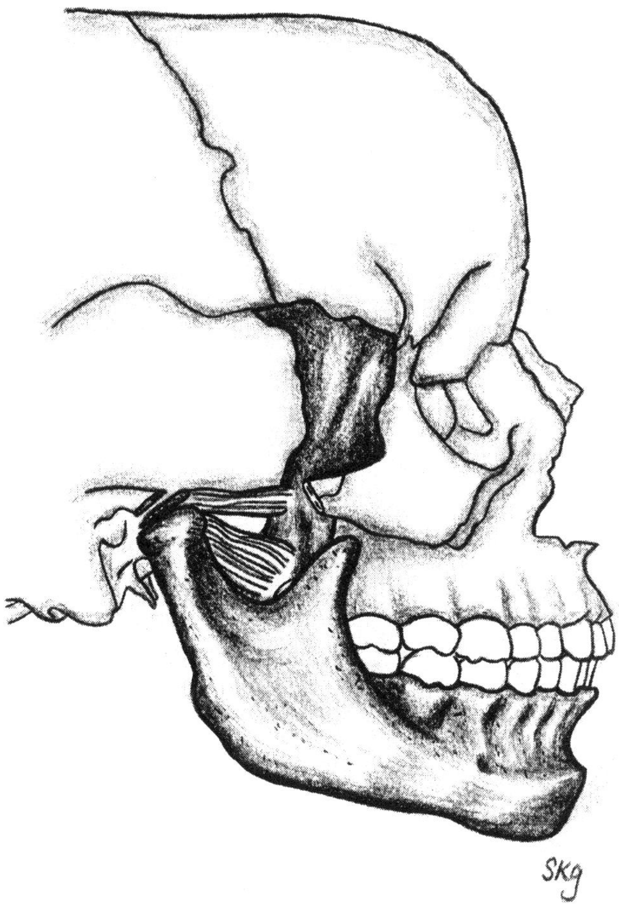

Origin:
Superior head: Sphenoid bone (greater wing, infratemporal crest and surface),
Inferior head: Mandible (lateral pterygoid plate & lateral surface)

Insertion:
Mandible (condyloid process, medial 1/2 to 2/3 of neck),
Temporomandibular joint (anterior margin of articular capsule and disc)

Action:
Protracts mandibular condyle and disc of TMJ joint forward while the mandibular head rotates on disc (aids opening of mouth). The lateral pterygoid, acting with the elevators of the mandible, protrudes the jaw, causing malocclusion of the teeth (the lower teeth project in front of the upper teeth). When the lateral and medial pterygoids on the same side act jointly, the mandible and the jaw (chin) rotate to the opposite side (in a chewing motion).

Innervation:
Trigeminal nerve (via cranial nerve V)

Relationship:
This is a short, thick muscle with two heads that runs posterolaterally to the mandibular condyle, neck, and disc of the TMJ. The fibers of the upper head are directed downward and laterally, while those of the lower head course horizontally. The muscle lies under the mandibular ramus. The lateral pterygoid is often associated with the pain of TMJ dysfunction. Severe referred pain, bruxism, crepitus, and tinnitus are also common complaints. Trigger point pain is referred to the cheek and TMJ area. Symptoms must be differentiated from those of sinusitis, tension headache, TMJ dysfunction, and trigeminal neuralgia.

PTERYGOID MEDIALIS (right)
Lateral View

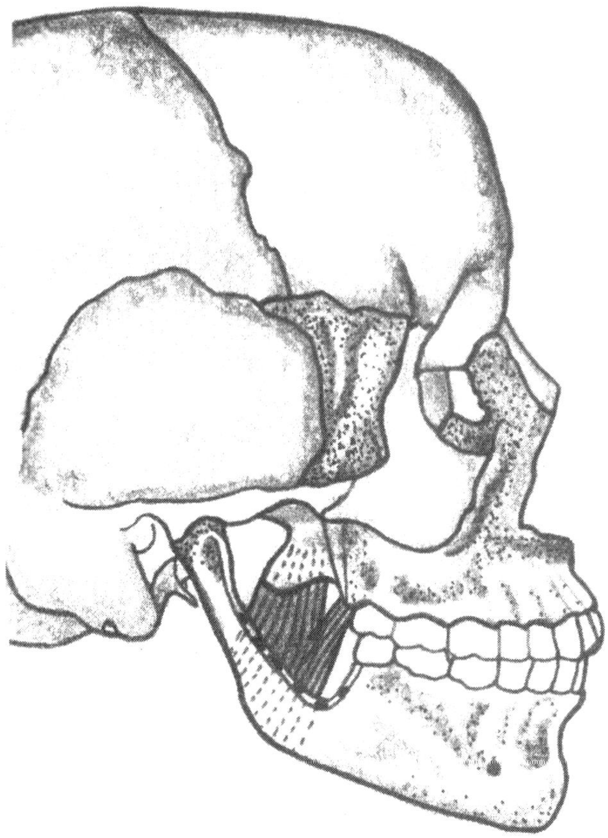

Origin:
Sphenoid bone (lateral pterygoid plate),
Palatine bone (grooved surface of pyramidal process),
Maxilla (tuberosity) joins in slip from lateral surfaces of pyramidal processes of palatine bone)

Insertion:
Mandible (medial surface of ramus via a strong tendon, reaching as high as mandibular foramen)

Action:
Elevates mandible to close jaws (biting),
Protrudes mandible (with lateral pterygoid),
Unilaterally the medial and lateral pterygoids together rotate the mandible forward and to the opposite side. This alternating motion is chewing.

Innervation:
Trigeminal nerve (via cranial nerve V)

Relationship:
This short, thick muscle occupies the position on the inner side of the mandibular ramus, whereas the masseter occupies the outer position. The medial pterygoid and masseter are situated to form a sling that suspends the mandible. This sling is a functional articulation in which the temporo-mandibular joint acts as a guide. As the mouth opens and closes, the mandible moves on a center of rotation established by the sling (spheno-mandibular ligament). Complaints from the medial pterygoid include local and referred pain (which is made worse by opening the mouth wide and chewing or by clenching the teeth), painful swallowing, and sore throat (without inflammation). Trigger point pain is referred to the posterior throat and into the ear. Symptoms must be differentiated from those of sinusitis, pharyngitis, and other dysfunctional muscles (SCM, digastrics, longus capitus, and longus colli).

TEMPORALIS (right)
Lateral View

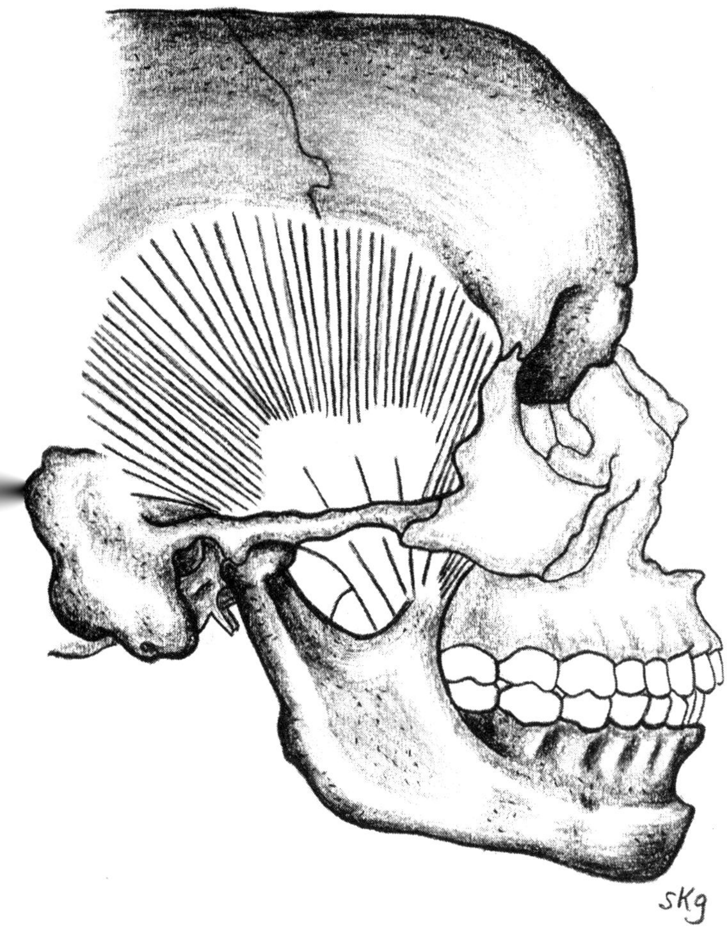

Origin:
Temporal bone (all of temporal fossa), Temporal fascia (deep surface)

Insertion:
Mandible (coronoid process, medial surface, apex. border; anterior border of ramus to 3rd molar)

Action:
Elevates mandible to close mouth and approximate teeth (biting motion), Retracts mandible (posterior fibers), Participates in lateral grinding motion

Innervation:
Trigeminal nerve (via cranial nerve V)

Relationship:
Temporalis is a broad muscle that radiates like a fan on the side of the head from most of the temporal fossa, converging downward to the coronoid process of the mandible. The descending fibers converge into a tendon that passes between the zygomatic arch and the cranial wall. The more anterior fibers descend vertically, but the more posterior the fibers run almost horizontally. The temporalis is responsible for maintaining the mandible in the closed position during the upright posture. Complaints from the temporalis commonly include headache, neckache, and tooth pain with hyper-sensitivity (hot, cold, and percussion). Symptoms may be aggravated by clenching, bruxism, direct trauma, immobilization, and therapeutic traction. Trigger point pain is referred to the eyebrow, upper teeth, and TMJ area. Symptoms must be differentiated from those of TMJ dysfunction, dental disease, tension headache, cervicogenic headache, and temporal arthritis.

MASSETER (right)
Lateral View

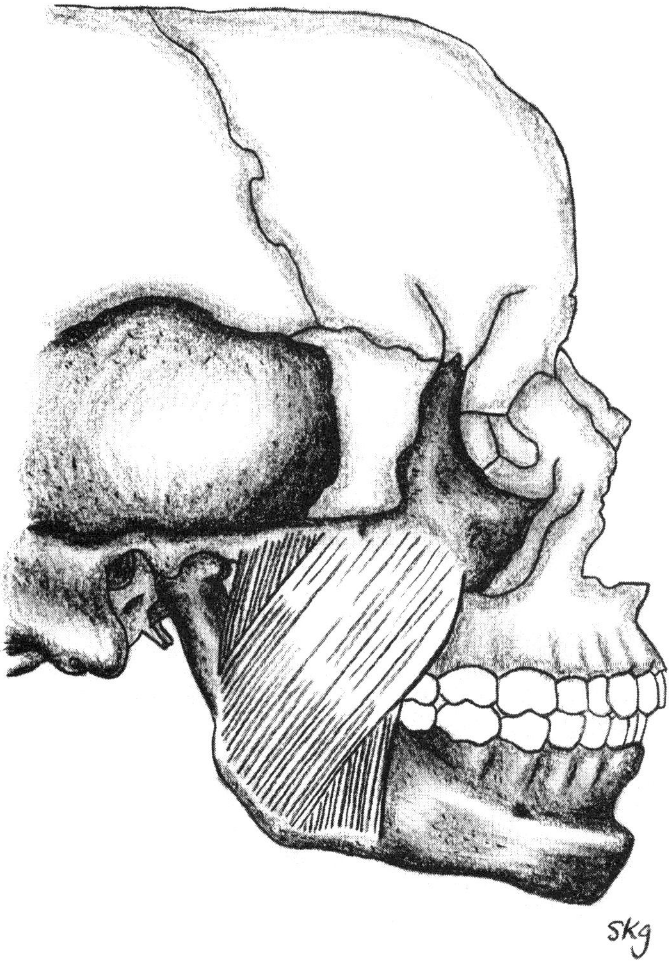

Origin:
Maxilla (zygomatic process via an aponeurosis),
Zygomatic bone (maxillary process and medial & inferior border of arch)

Insertion:
Mandible (angle and lower half of lateral surface of ramus) and coronoid process

Action:
Elevates the mandible (occlusion of the teeth in mastication), up and down biting motion

Innervation:
Trigeminal nerve (via cranial nerve V)

Relationship:
The masseter is a thick muscle connecting the upper and lower jaws and consisting of three layers that blend. The superficial layer descends backward to the angle of the mandible and the lower mandibular ramus. The muscle is easily palpable and lies under the parotid gland. Posteriorly, the anterior margin overlies the buccinator. Complaints associated with the masseter are local pain, restriction in opening the jaw, and unilateral tinnitus. These complaints are caused or aggravated by direct trauma, bruxism, chronic overuse, and chronic clenching. Trigger point pain is referred to the eyebrow, ear, anterior maxilla & mandible, and teeth. Symptoms must be differentiated from those of tinnitus of neurological origin, TMJ dysfunction, dental disease, tension headache, cervicogenic headache, and adjacent head and neck muscles.

ANTERIOR MUSCLES of the NECK
Suprahyoid

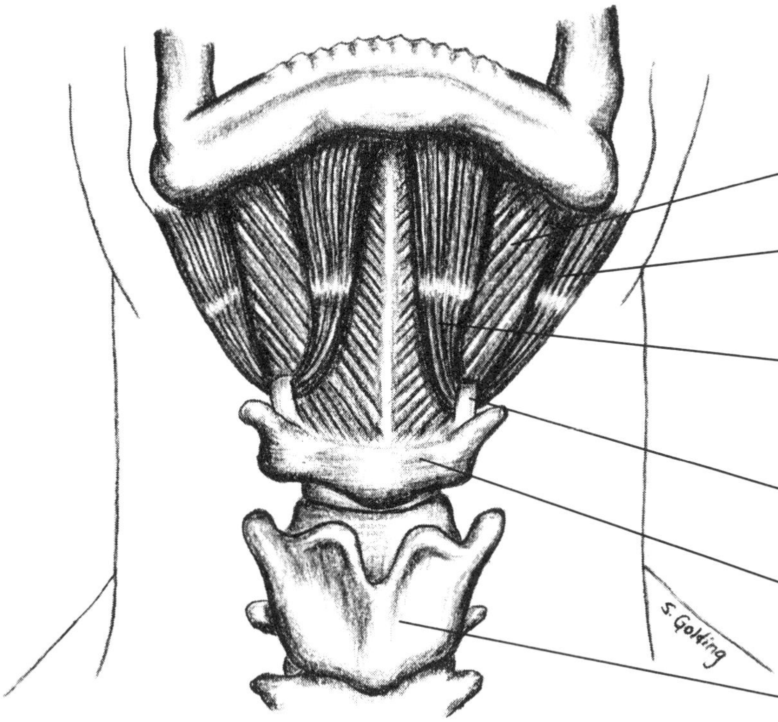

Mylohyoids
Origin:Inner surface of Mandible
Insertion: Superior surface (body) of Hyoid
Action: Elevates Hyoid (or lowers mandible)
Innervation: Cranial Nerve V (& alveolar)

Digastrics
Origin: Mandible & Mastoid process (posterior)
Insertion: Body of Hyoid via Fibrous loop
Action: Elevates Hyoid (or lowers mandible)
Innervation: Cranial Nerves V & VII

Geniohyoids (shown on next page)
Origin:Inner surface of Mandible
Insertion: Superior surface (body) of Hyoid
Action: Elevates Hyoid (or lowers mandible)
Innervation: Cranial Nerve XII (via C-1)

Stylohyoid (not illustrated)
Origin: Styloid process of temporal bone
Insertion: Superior surface (body) of Hyoid
Action: Elevates Hyoid
Innervation: Cranial Nerve VII

ANTERIOR MUSCLES of the NECK
Infrahyoid

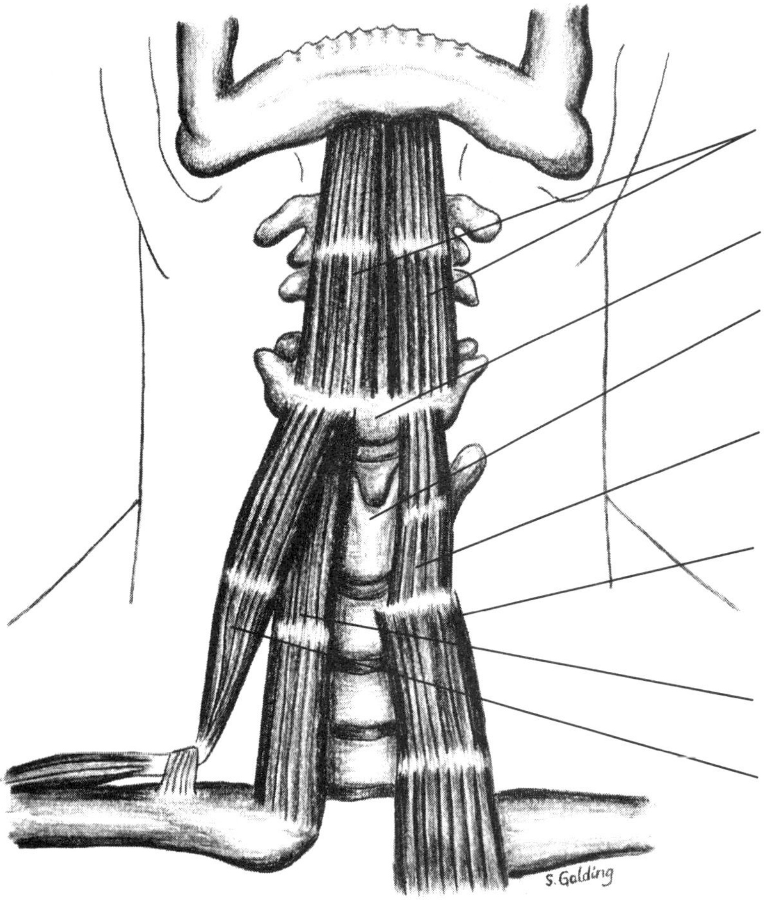

Thyrohyoid
Origin: Thyroid Cartilage
Insertion: Greater horn of Hyoid
Action: Stabilizes Hyoid during deglutition
Innervation: Cranial Nerve XII & C-1, C-2

Sternothyoid
Origin: Sternum (superior/posterior manubrium)
Insertion: Surface of Thyroid cartilage
Action: Depresses thyroid cartilage
Innervation: C-1 thru C-3

Sternohyoid
Origin:Sternum & posterior/medial Clavicle
Insertion: Inferior surface (body) of Hyoid
Action: Stabilizes Hyoid during deglutition
Innervation: C-1 thru C-3

Omohyoid
Origin: Upper border of scapula (near notch)
Insertion: Anterior/Inferior surface of Hyoid
Action: Stabilizes Hyoid during deglutition
Innervation: C-1 thru C-3

INTERVERTEBRAL JOINTS & DISCS
(left) Lateral View

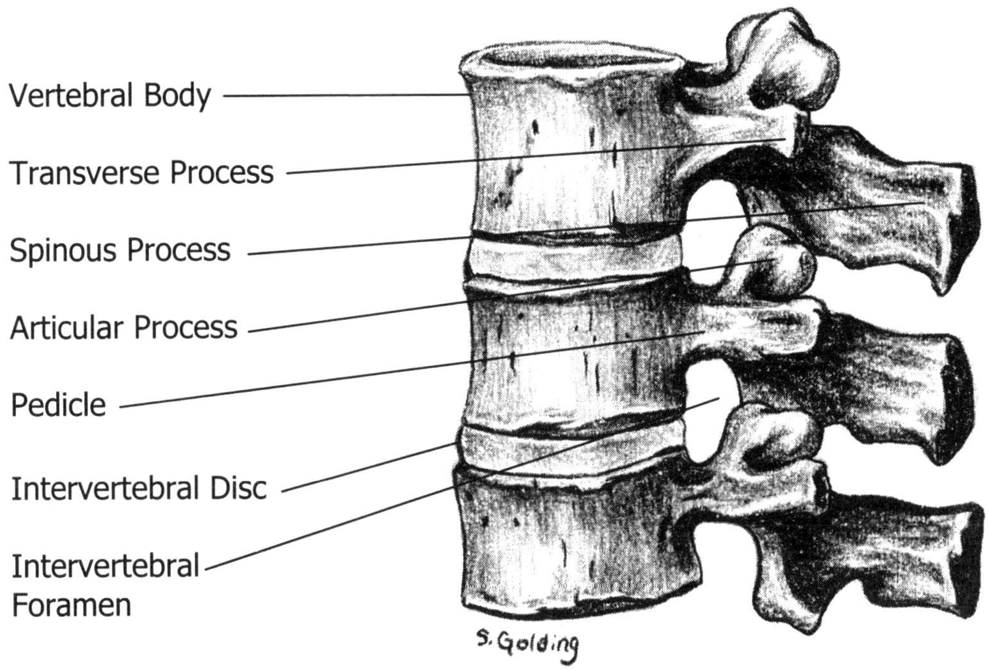

INTERVERTEBRAL DISC
Superior View

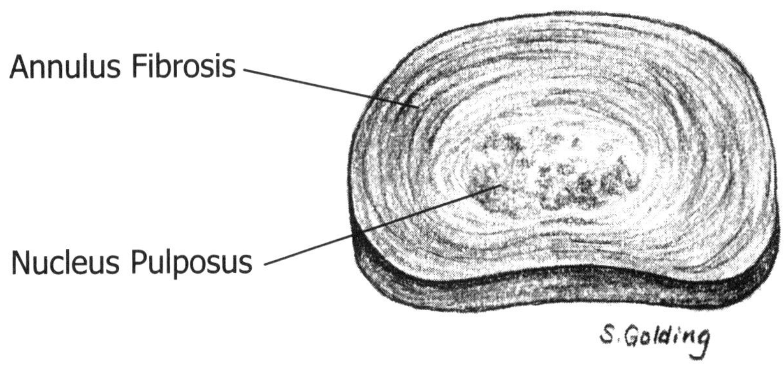

Intervertebral articulations (except atlas-axis) are characterized by amphiarthrotic or cartilaginous joints, which utilize discs between vertebral bodies and diarthrotic or synovial, gliding (facet) type joints between the articular processes of each vertebra. The intervertebral disc has an outer fibrous ring and an inner soft, elastic nucleus. Each disc corresponds in surface area and thickness to its adjacent vertebrae. Discs are nearly avascular, yet absorb significant amounts of water from surrounding tissues. With age, discs become less hydrated, less flexible, more fibrous, and more susceptible to trauma.

SACRO ILLIAC JOINT& LIGAMENTS (right)
Posterior View

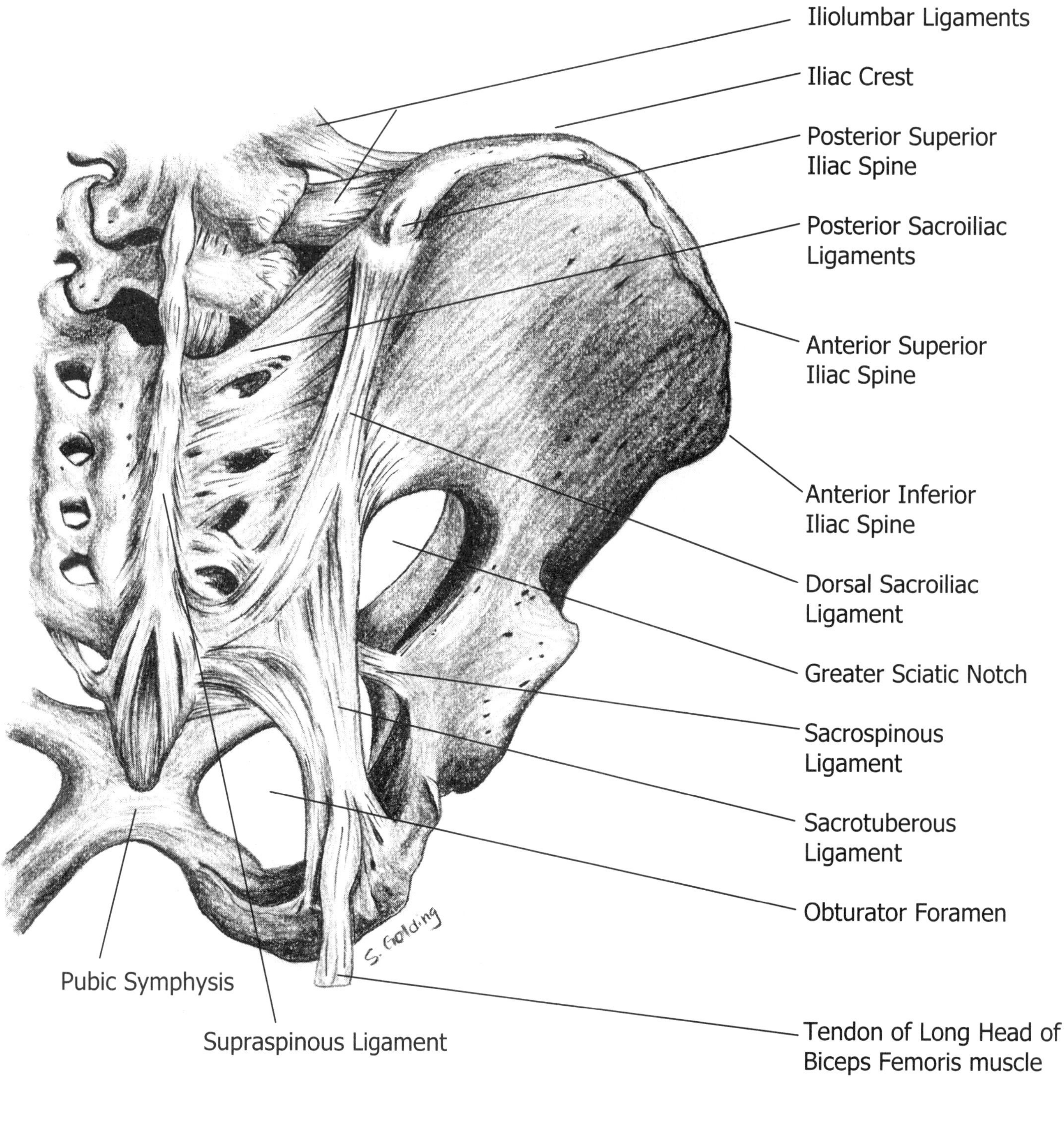

SHOULDER JOINT & LIGAMENTS (right)
Anterior View

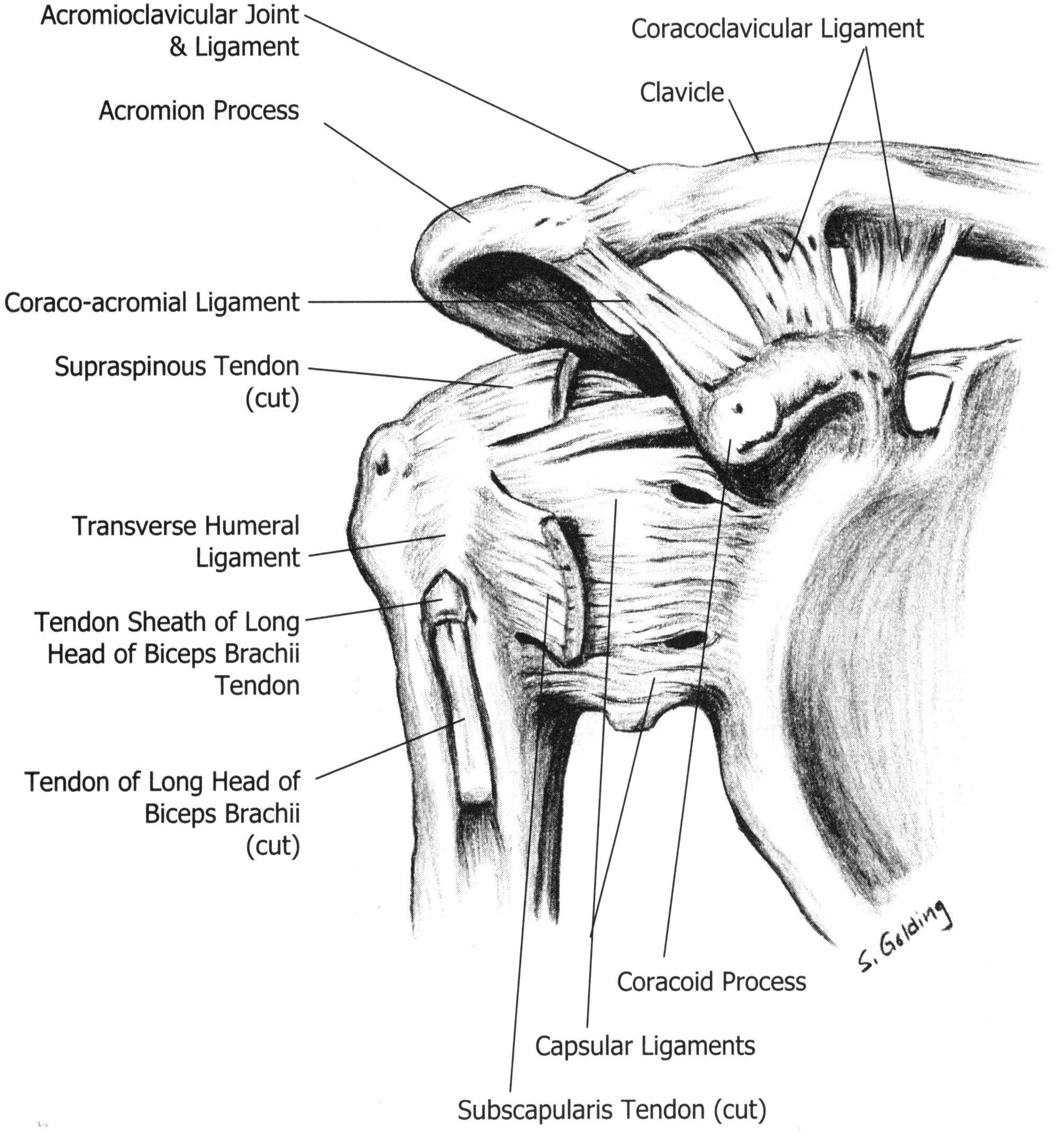

ELBOW JOINT (right) & LIGAMENTS
Medial View

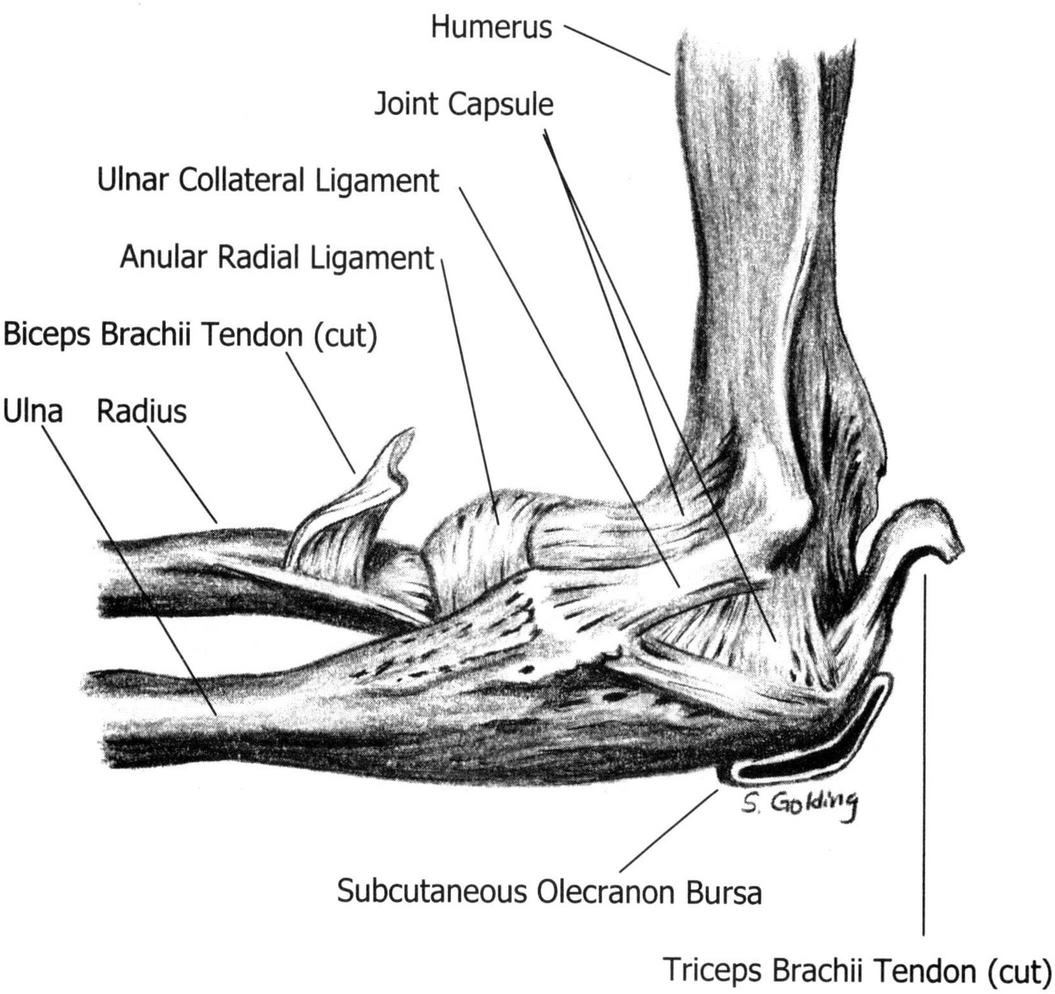

ELBOW JOINT (right) & LIGAMENTS
Lateral view

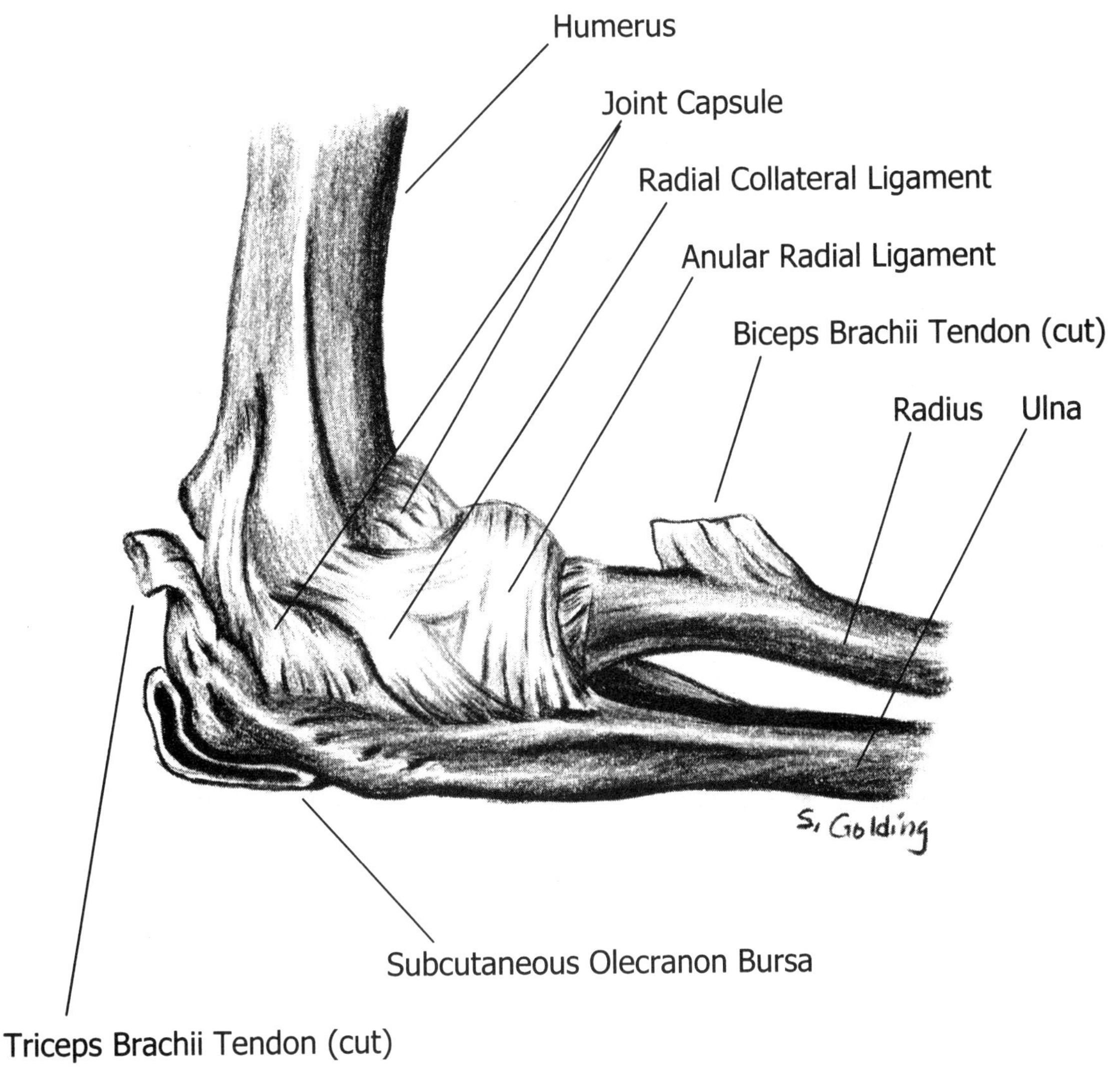

WRIST JOINT (right) with VESSELS, TENDONS & LIGAMENTS
Anterior or Palmar View

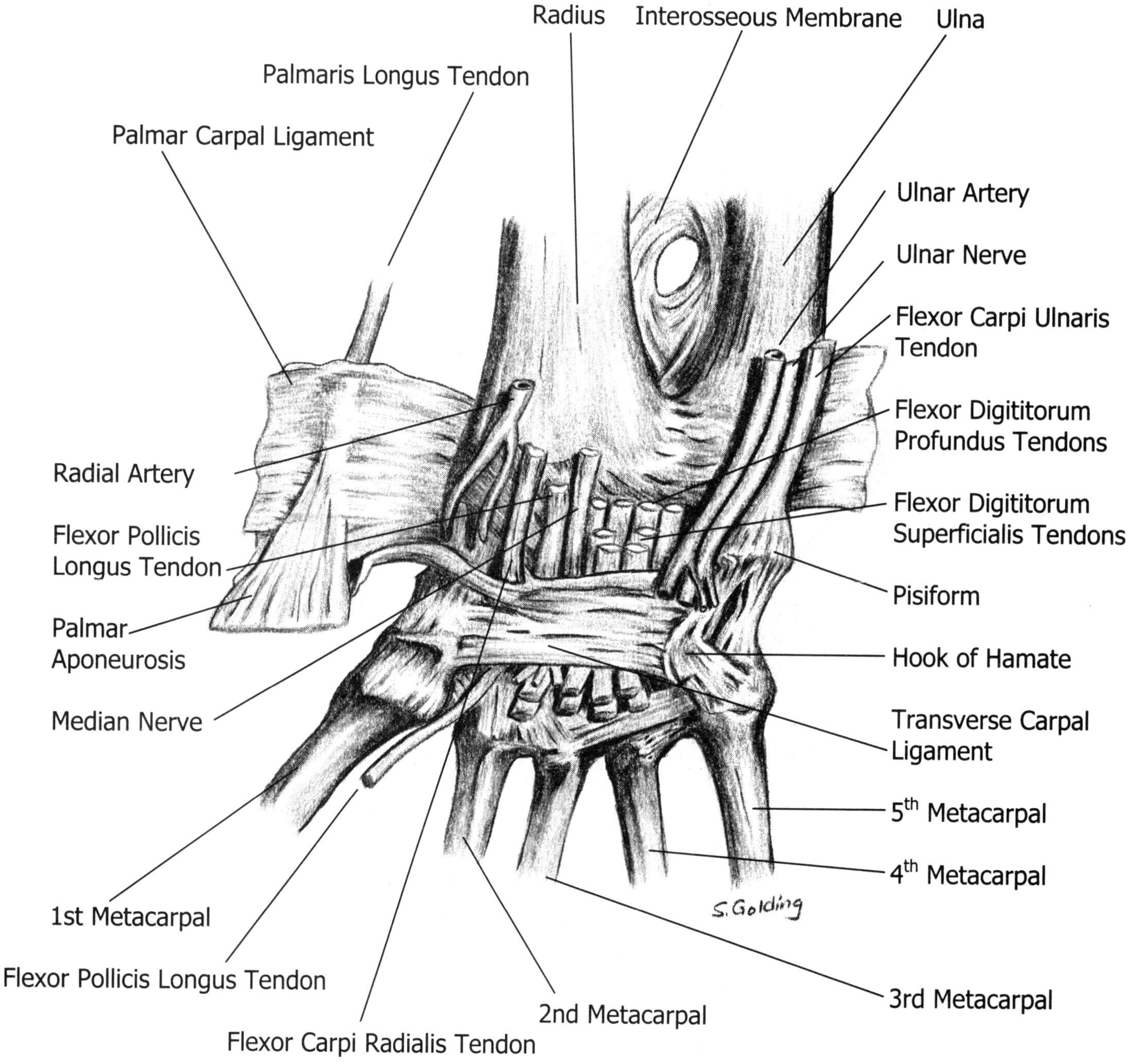

KNEE JOINT (right)
in flexion with
LIGAMENTS & MENISCUS
Anterior View

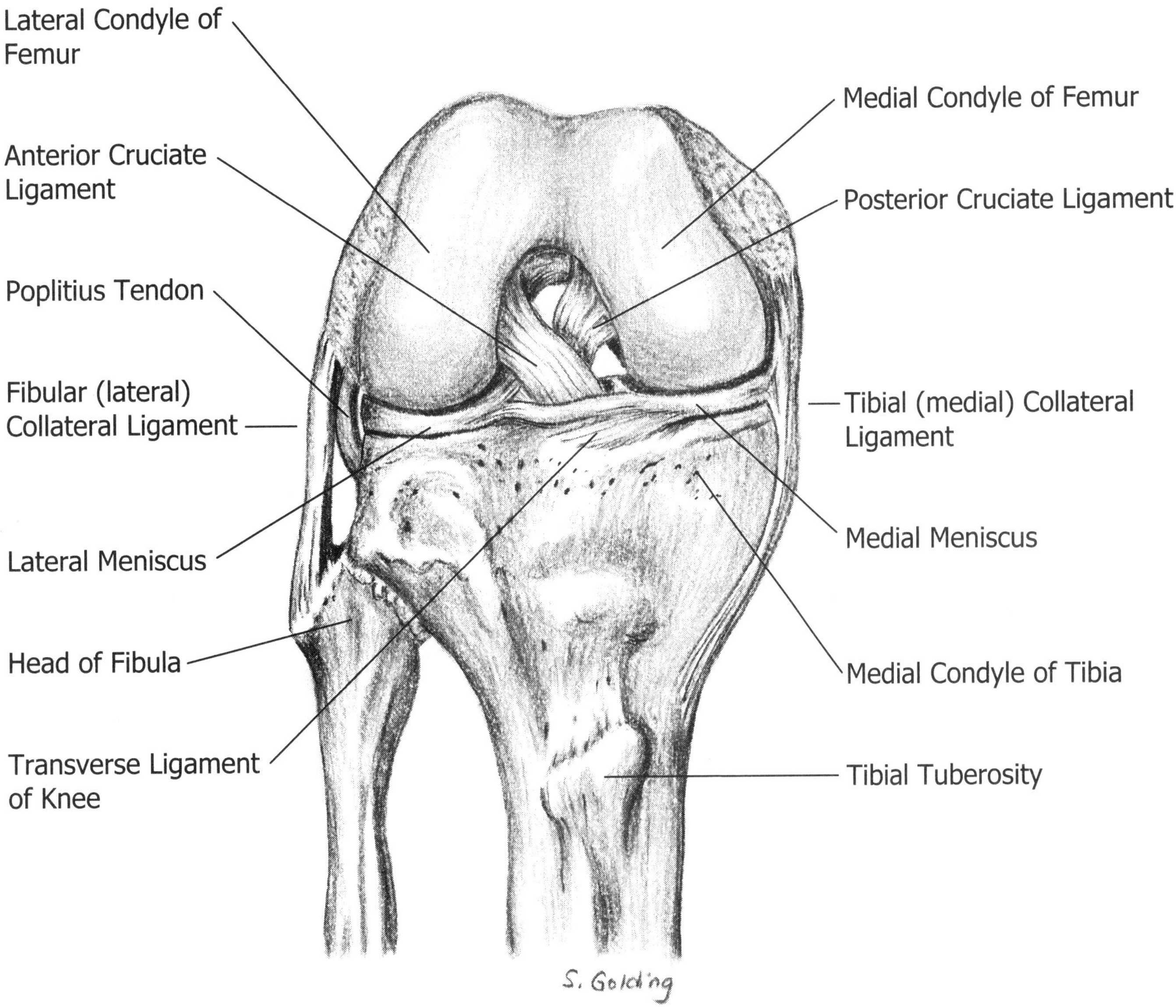

KNEE JOINT (right) with LIGAMENTS & MENISCUS
Posterior View

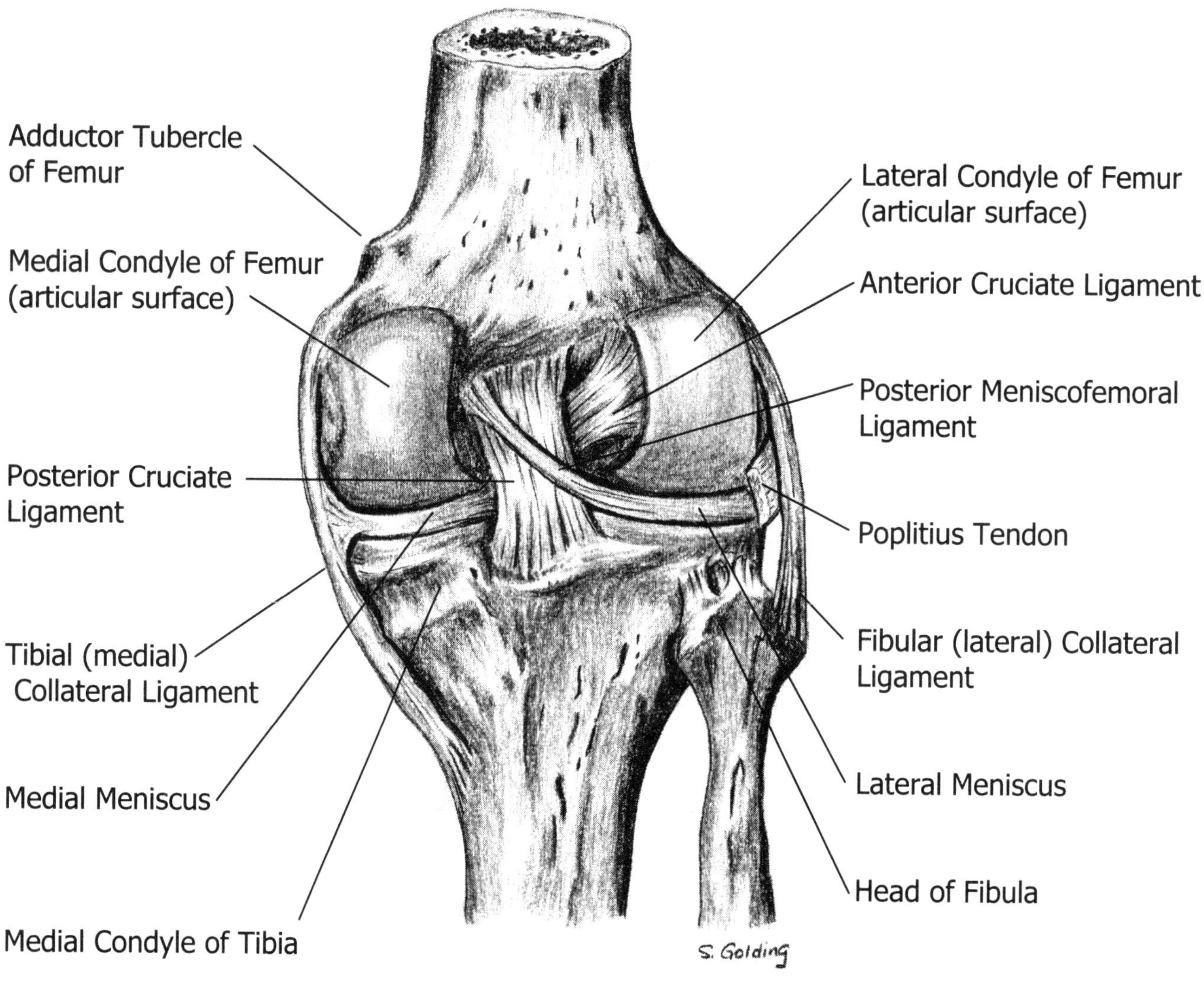

ANKLE JOINT (right) & LIGAMENTS
Medial View

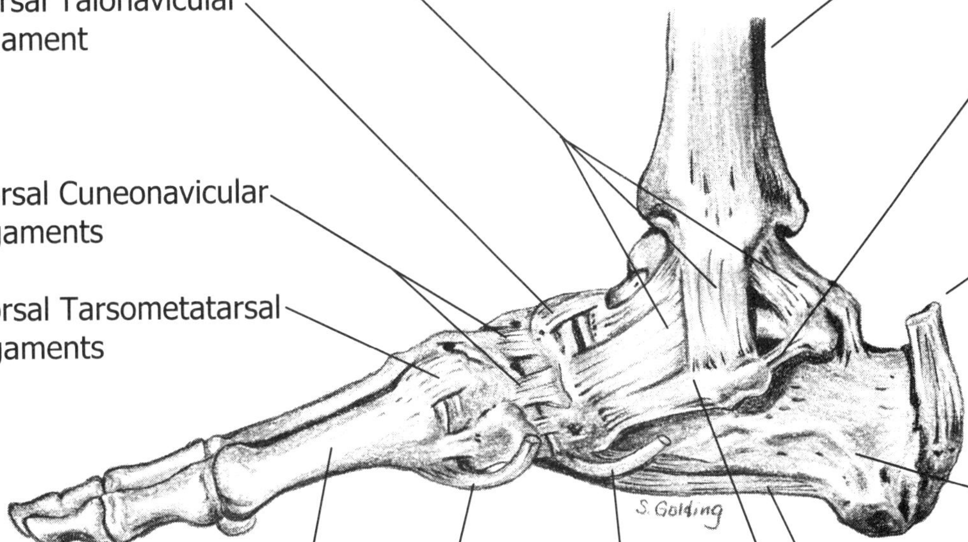

ANKLE JOINT (right) & LIGAMENTS
Lateral View

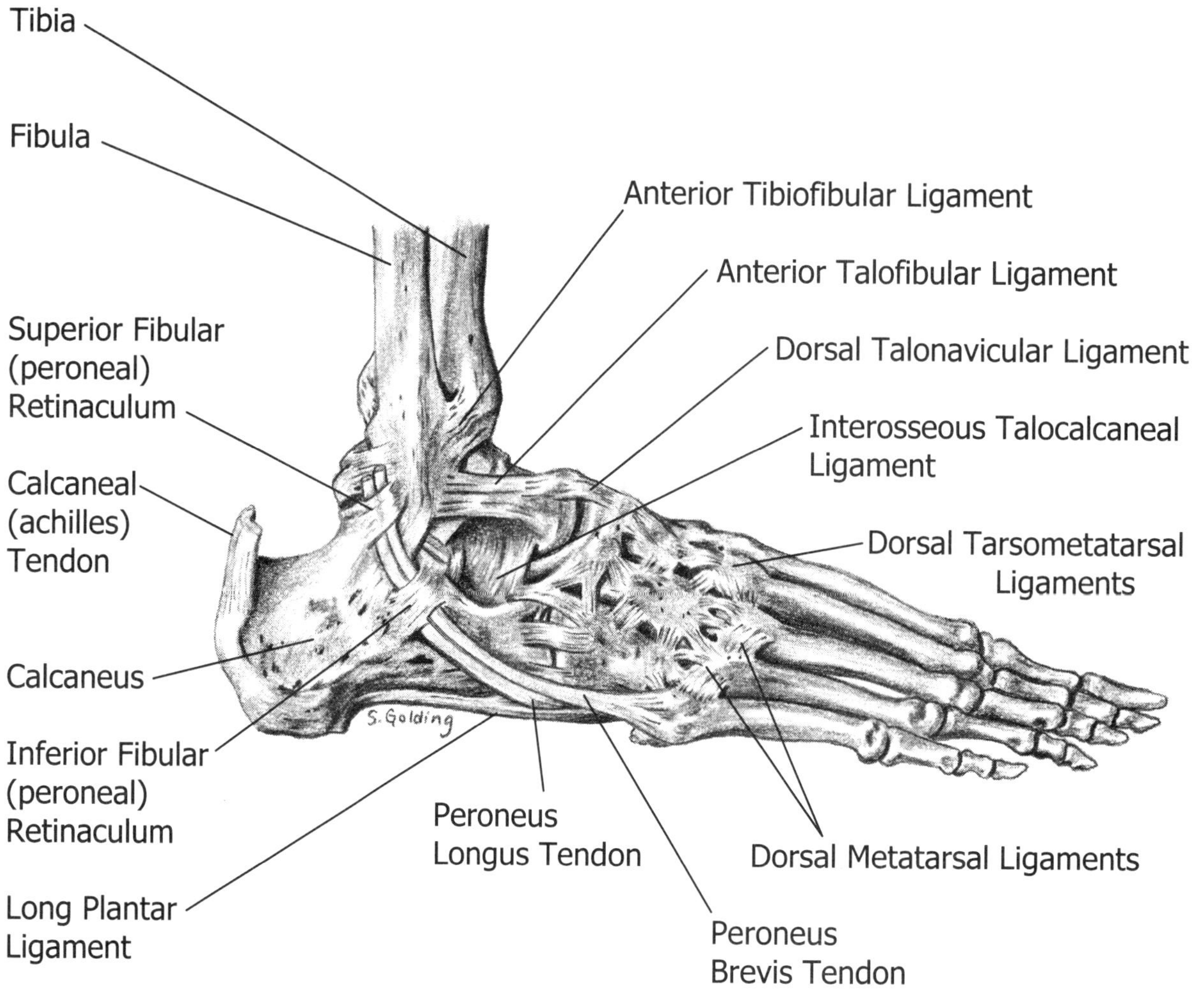

BRACHIAL PLEXES (right) Anterior View

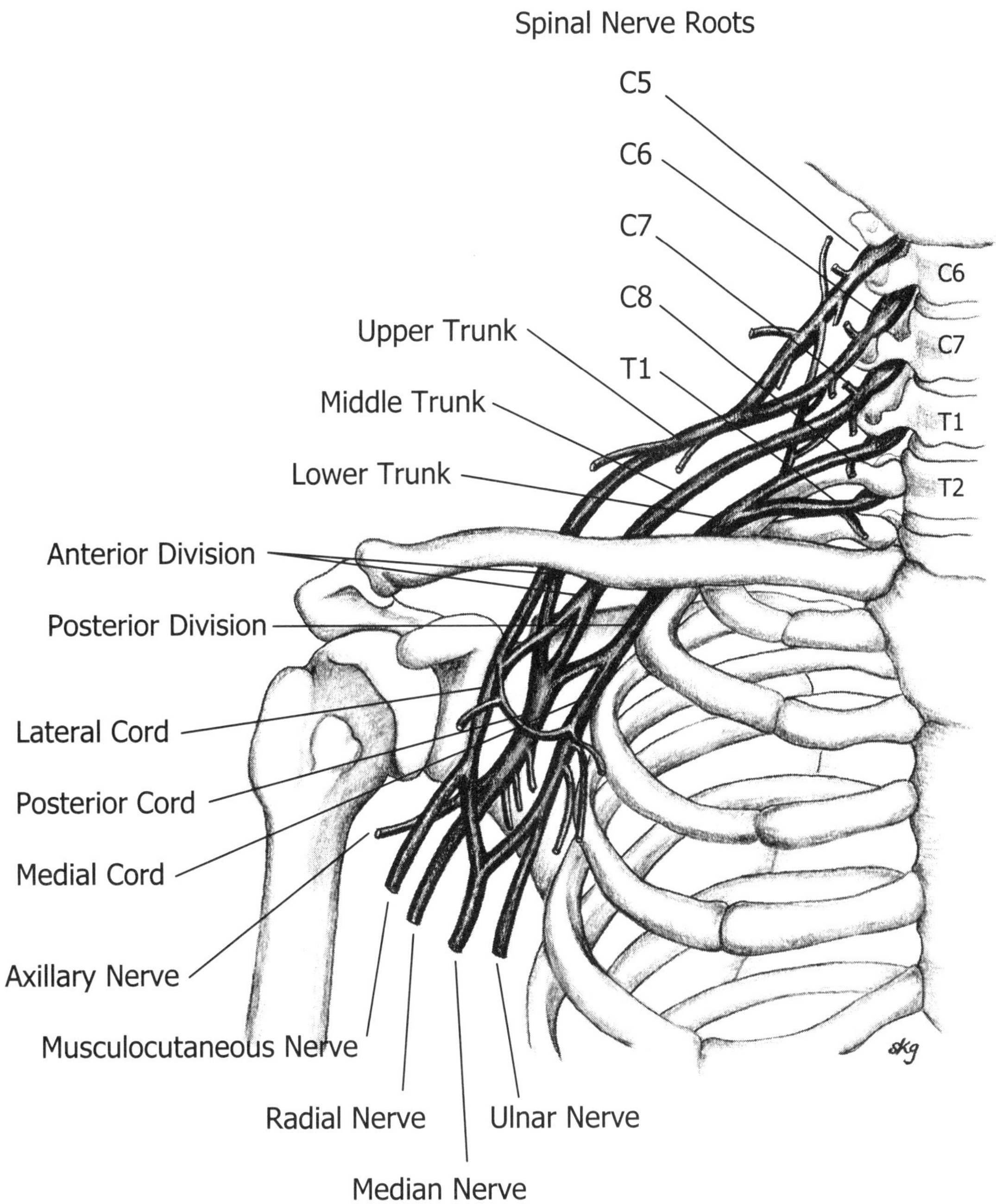

The brachial plexus is formed by the union of the anterior (ventral) rami of spinal nerves C5 thru C8 and T1. It extends laterally and inferiorly on either side of cervical vertebrae C4 thru C7 and T1. The brachial plexus in found in the posterior triangle, deep to platisma. Emerging from between the anterior and middle scalenes, it passes deep to both the clavicle and pectoralis minor on its route to the axilla. The axillary artery accompanies the brachial plexus as is passes under the clavicle and pectoralis minor and over the rib cage. The brachial nerve may be compromised by hypertonic scalene or pectoralis minor muscles, the presence of a cervical rib, or a number of other causes.

LUMBOSACRAL PLEXUS (right) Anterior View

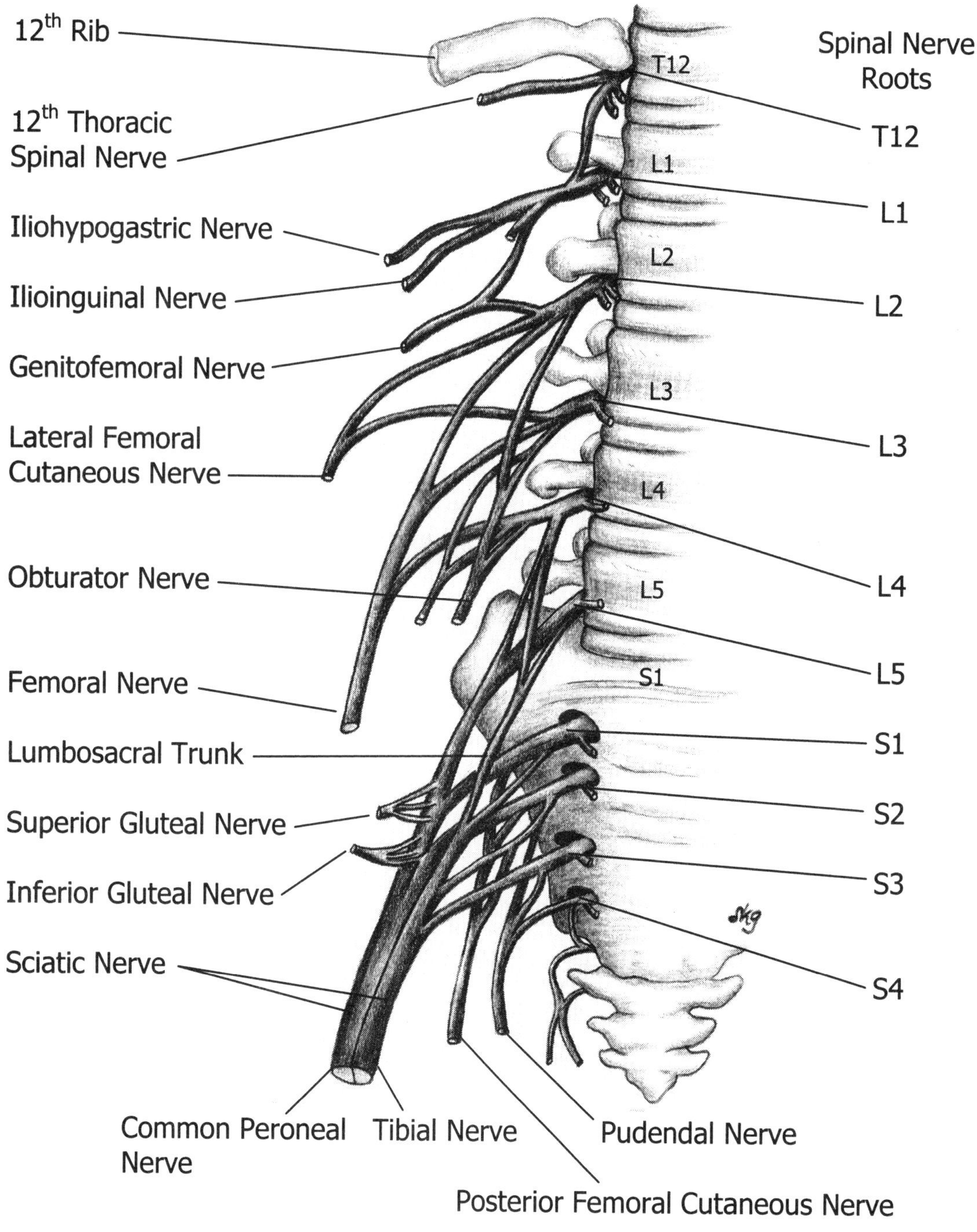

The lumbosacral plexus is formed by the union of the anterior (ventral) rami of spinal nerves L1 thru S4. It extends laterally and inferiorly on either side of the lumbar vertebrae and anteriorly to the sacrum. The lumbosacral plexus passes posterior to psoas major and anterior to quadratus lumborum. The sciatic nerve (largest in the body) arises from L5 - S4 portion of the plexus and travels deep to the piriformis muscle (which when hypertonic, may cause compression of the nerve and result in sciatica, ie: pain along the course of the sciatic nerve).

INDEX

WORKS CITED

Altug, Ziya & Hoffman, Janet, and Martin, Jerome. *Manual of Clinical Exercise Testing, Prescription and Rehabilitation*. Norwalk: Appleton & Lange, 1993.

Art & Science Studio, website: http://www artandsciencestudio org/2005_anatomy.

Cailliet, Rene. *Soft Tissue Pain and Disability*. 2nd Edition. Philadelphia: F A Davis Company, 1988.

Calais-Germain, Blandine. *Anatomy of Movement.* Seattle: Eastland Press, 1993.

Clemente, Carmine D. *Anatomy; A Regional Atlas of the Human Body*. 3rd Edition. Baltimore: Urban & Schwrzenberg, l987.

Greenman, Phillip. *Principles of Manual Medicine*. 2nd Edition. Baltimore: Lippincott, Williams & Wilkins, 1996.

Guyton, Arthur C and Hall, John E. *Textbook of Medical Physiology*. Philadelphia: W B Saunders Co, 1996.

Hole, John W Jr. *Essentials of Human Anatomy & Physiology.* Dubuque: Wm C Brown Publishers, 1992.

Hoppenfeld, Stanley. *Physical Examination of the Spine and Extremities*. New York: Appleton-Century-Crafts, l976.

Juhan, Deane; *Job's Body-A Handbook for Bodywork*, Barringtown, New York: Station Hill Press, l987

Kulund, M D, and Daniel, N. *The Injured Athlete*. Philadelphia: J B Lippincott Co, 1988.

Mayo Clinic, Educational website: http://www mayo edu/medlib_exhibit/18th.

Anderson, D. et al. *Mosby's Medical Nursing and Allied Health Dictionary*. St, Louis: Harcourt Health Sciences Inc., 2002.

National Library of Medicine, National Institute of Health, website: http://www.nlm.nih. gov/exhibition/dreamanatomy.

Netter, Frank, H. *Atlas of Human Anatomy*. 2nd Edition. Summit, NJ: Novartis Medical Education, 1997.

Robbins. *Pathologic Basis of Disease*. 6th Edition. New York: Saunders, 2000.

Roy, Steven, and Irvin, Richard. *Sports Medicine, Prevention, Evaluation, Management, and Rehabilitation*. Prentice Hall Publishing Inc, 1983.

Stone, Robert J and Stone, Judith A. *Atlas of the Skeletal Muscles*. Dubuque: Wm C Brown, 1990.

Thomas, C L, ed. *Taber's Cyclopedic Medical Dictionary*. 16th Edition. Philadelphia: F A Davis Co, l989.

Thompson & Floyd. *Manual of Structural Kinesiology*. 14th Edition. Boston: McGraw-Hill, 2001.

Tortora and Grabowski. *Principles of Anatomy and Physiology*. 10th Edition. New York: John Wiley & Sons, 2003.

Travell & Simons. *Myofascial Pain and Dysfunction: The Trigger Point Manual, Volumes I and II*. Philadelphia: William & Wilkins, l992.

University of Dayton website: http://www.udayton edu/~hume/Galen/galen.

University of Glasgow, Special Collections website: http://www.special.lib.gla.ac.uk/anatomy.

University of Leeds website: http://www.leeds.ac.uk/chb/lectures/anatomy1.

Whiting, William C, and Zernicke, Ronald F. *Biomechanics of Musculoskeletal Injury*. Champaign: Human Kinetics, 1998.

Williams, Peter. *Gray's Anatomy*. 38th Edition. Philadelphia: W B Saunders Co, 1999.

NOTES

Virtuvian Man - from the sketchbook of Da Vinci

NOTES

Muscles - from the sketchbook of Da Vinci

NOTES

Head & Neck - from the sketchbook of Da Vinci

NOTES

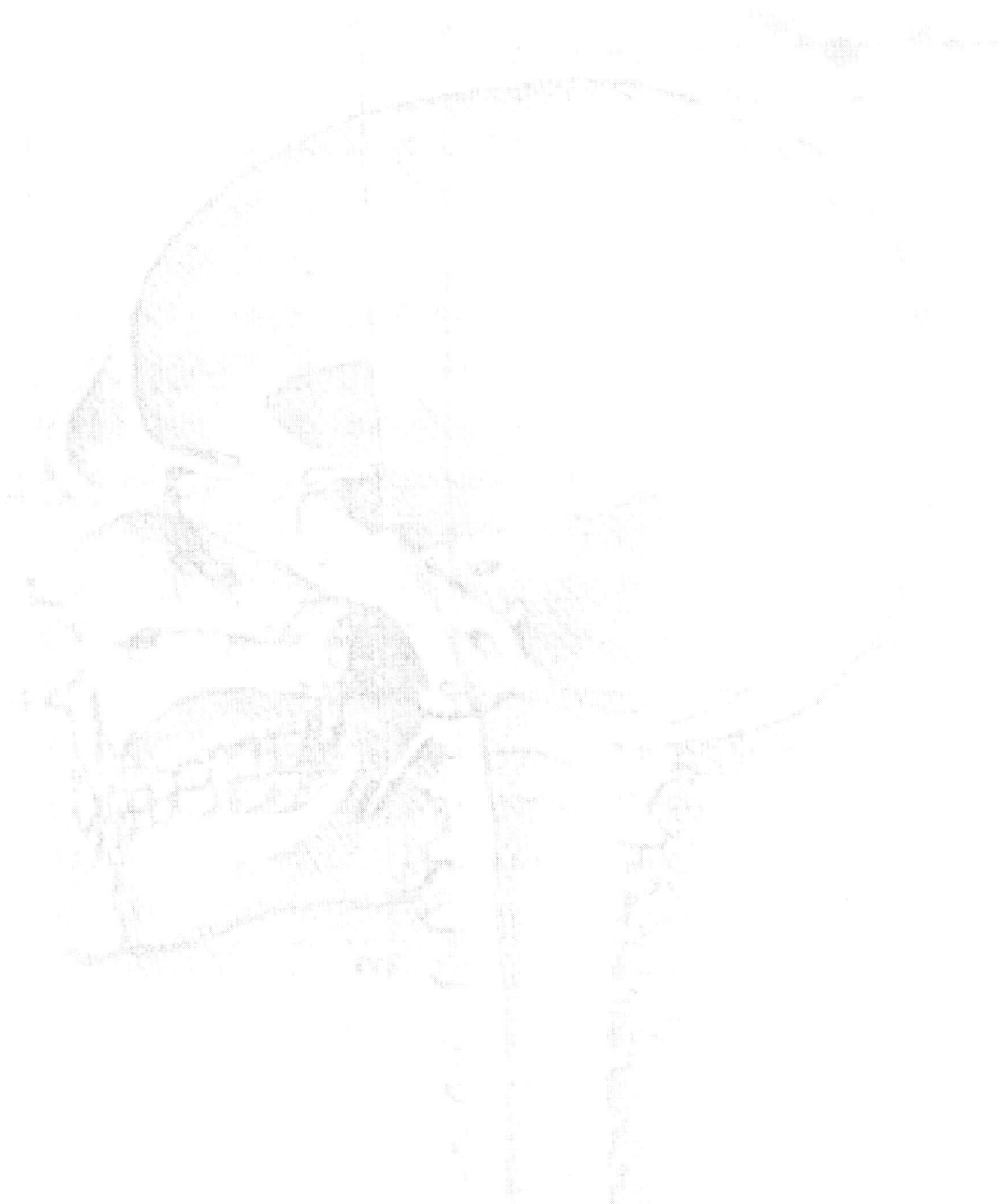

Skull - from the sketchbook of Da Vinci

NOTES

Arms - from the sketchbook of Da Vinci

NOTES

Fetus - from the sketchbook of Da Vinci